T0138387

Inventing Chemistry

synthesis

A series in the history of chemistry, broadly construed, edited by Angela N. H. Creager, John E. Lesch, Stuart W. Leslie, Lawrence M. Principe, Alan Rocke, E. C. Spary, and Audra J. Wolfe, in partnership with the Chemical Heritage Foundation.

Inventing Chemistry

Herman Boerhaave and the Reform of the Chemical Arts

JOHN C. POWERS

The University of Chicago Press

CHICAGO AND LONDON

JOHN C. POWERS is collateral assistant professor in the Department of History
and assistant director of the Science, Technology, and Society Program
at Virginia Commonwealth University.

The University of Chicago Press, Chicago 60637
The University of Chicago Press, Ltd., London

Printed in the United States of America

21 20 19 18 17 16 15 14 13 12 1 2 3 4 5

ISBN-13: 978-0-226-67760-6 (cloth)
ISBN-10: 0-226-67760-5 (cloth)

Library of Congress Cataloging-in-Publication Data

Powers, John C., 1968–
Inventing chemistry : Herman Boerhaave and the reform of the chemical arts / John C. Powers.
p. cm.
Includes bibliographical references and index.
ISBN-13: 978-0-226-67760-6 (cloth : alkaline paper)
ISBN-10: 0-226-67760-5 (cloth : alkaline paper) 1. Boerhaave, Herman, 1668–1738.
2. Chemistry—History. 3. Chemistry—Study and teaching—Netherlands—Leiden—History—
18th century. I. Title.
QD15.P69 2012
540.92—dc23 2011033952

♾ This paper meets the requirements of ANSI/NISO Z39.48-1992 (Permanence of Paper).

Contents

Acknowledgments

This book is the culmination of many years of work beginning in my graduate student days at Indiana University in the 1990s. Along the way, many people have given me support, assistance, advice, and encouragement, and without them I could not have completed this project.

First, I thank my advisors and friends at the Department of History and Philosophy of Science at Indiana University, under whose guidance and encouragement this project was born. My early approach to Boerhaave was shaped by my experience in the history, sociology, and history and philosophy of science communities, which included Richard Sorrenson, William Newman, Domenico Bertoloni Meli, Ann Carmichael, Tom Gieryn, Ed Grant, Noretta Koertge, James Capshew, Patrick Catt, Brad Hume, Eric Winsberg, Elizabeth Green-Musselman, Peter Ramberg, Judy Johns Schloegel, and Peter Sobol.

The heart of this project has been the study of manuscripts at the Military-Medicine Academy (VMA) in St. Petersburg, Russia, the British Library, and the Universiteitsbibliotheek Leiden. I thank the library staff at each of these institutions for their assistance with my research. I especially thank Galina Mikhailovna, the "keeper of Germanic manuscripts" at the VMA for her extra efforts in helping me navigate the Boerhaave the archive. My gratitude also goes out to the Institute for the History of Science and Technology in St. Petersburg, which sponsored my research trip to Russia, and whose members, including Daniel Alexandrov, Eduard Kolchinsky, Sergei Trokachev, Julia Lajus, Lloyd Ackert, and Mikhail Konashev, helped me locate and gain access to the materials at the VMA. Finally, I thank Olga Melnichenko, who opened her home to me while I was in St. Petersburg.

Several institutions have supported me during the writing of this book. The Chemical Heritage Foundation took me on as a research assistant in the early stages of this project and, then, had me back as a researcher. I thank Arnold Thackray and Mary Ellen Bowden for their help and encouragement, and James Voelkel for his research assistance. I also thank Judy Goodstein and the archives of the California Institute of Technology, Steve Hilgartner and the Department of Science and Technology Studies at Cornell University, and all of my friends at Sarah Lawrence College. Finally, I thank my colleagues in the Department of History and in the Science, Technology, and Society Program at Virginia Commonwealth University, especially Leigh Ann Craig, Bernard Moitt, Marian Moser Jones, and Wanda Clary.

Most of the research for this project was made possible by support from the National Science Foundation, Program in Science, Technology and Society. The NSF supported this project financially on two occasions—with a dissertation grant (no. 9616667) in 1998 and a Scholar's Award (no. 0350017) in 2003.

In addition to the assistance given by institutions, I have benefited from conversations, suggestions, and paper exchanges with numerous individuals. Among these are Lawrence Principe, Jan Golinski, Frederic Holmes, Hasok Chang, Lissa Roberts, Mordechi Feingold, Seymour Mauskopf, Pamela Smith, Kevin Chang, Matthew Eddy, Mi Gyung Kim, Ursula Klein, John Detloff, Jonathan Simon, Craig Martin, and Evan Ragland. I especially thank the editor of the Synthesis series, Audra Wolfe, who read and commented on two early versions of the manuscript. I also thank three anonymous reviewers for their comments and encouragement.

Last, I thank my family for their support through the long and difficult process that is writing a book. I thank my parents, Charles and Kathleen Powers, for their unwavering faith and constant encouragement. I thank my children, Sam and Grace, for their patience with me, especially during the few days before each deadline. My special thanks and love go to Karen, who is my light and my strength. She has labored through the production of this book, if in different ways, as much as I have.

INTRODUCTION

BOERHAAVE IN THE HISTORY OF CHEMISTRY

Herman Boerhaave (1668–1738) devoted his professional life to improving and expanding the medical curriculum at the University of Leiden. During his thirty-seven-year academic career (1701–38), he held the chairs of botany and chemistry, taught the *institutiones medicae* course (the institutes of medicine, i.e., the basic theory course) and the medical praxis course (i.e., the application of theory to the treatment of specific diseases), conducted the clinical practicum, and gave private lectures on subjects ranging from anatomy to mixed mathematics. When he was named chair of botany and medicine in 1708, he took over Leiden's botanical garden, reorganized and expanded the plant and natural history collections, and published two botanical catalogs of the garden's holdings. He became chair of chemistry in 1718 and revamped the university's chemical laboratory.[1] Ultimately, he published a textbook in each of the five medical fields represented by chairs on the Leiden medical faculty.[2] By the 1720s Boerhaave virtually controlled the content of medical teaching at Leiden. All medical courses were taught by his allies or former students, and his method of medical education shaped the curriculum. He continued until the last year of his life to devise new, specialized courses, update and expand Leiden's basic medical courses, and publish revised editions of his textbooks.

Boerhaave's steady stream of publications and numerous students made him the most revered medical professor of the early eighteenth century. His student Albrecht von Haller called him "praeceptor communis Europae," reflecting his vast influence over medical education. A. L. Turner estimated that more than 1,900 students matriculated at the Leiden medical faculty during Boerhaave's years as a professor.[3] Boerhaave personally acted as "promoter" for 178 of those students during the ceremonial thesis disputation conducted as part of Leiden's graduation exercises.[4] His medical teachings were widely disseminated, both through his writings and through his students' attempts to reproduce the "Leiden model" of medical education in Edinburgh, Vienna, Göttingen, and, a generation later, Philadelphia. In fact, each of the five original professors on the medical faculty at Edinburgh had matriculated at Leiden, and to establish the credibility of the school, they advertised the fact that they followed Boerhaave's methods and texts.[5]

Boerhaave's work in chemistry had an even greater impact on shaping that field. He first began to teach chemistry as a private lecturer in 1702 and served as professor of chemistry on the Leiden medical faculty from 1718 to 1729. During his tenure, he completely revamped the manner in which chemistry was taught at Leiden, introducing both a new pedagogical method and theoretical framework for his chemistry courses. His textbook, *Elementa Chemiae* (1732), which derived from these courses, was just as popular as his medical works. The *Elementa* underwent at least forty separate printings through 1791, including translations from Latin into English, French, German, and Russian, even though Boerhaave himself was responsible for only one Latin and one English edition.[6] His chemical work, which has yet to be analyzed fully, exerted a strong influence on the development of the eighteenth-century chemical arts and sciences. For example, recent work has shown that prominent chemists in France and Scotland adopted Boerhaave's pedagogical and theoretical frameworks to structure their chemistry courses.[7] His work on chemical thermometry served as the foundation for the introduction of thermometry into brewing.[8] Similarly, Swedish chemists deployed his chemistry to reshape their chemical and mineralogical practices, abandoning the Paracelsian traditions that had previously been in place.[9]

Despite Boerhaave's acknowledged importance for eighteenth-century chemistry, historians have had a difficult time placing his work within the historical development of chemistry. This difficulty was caused in large part by the fact that Boerhaave was primarily a pedagogue, organizer, and system builder. By contrast, the history of science has traditionally been written as a history of theoretical innovation, with science pedagogy—textbooks and courses—depicted as vehicles for the transmission of the received view. Thomas Kuhn, for example, famously argued that textbooks (and science education, generally) were essential for building and maintaining a community of scientific practitioners in a given field, but that these books transmitted the "normal" science view. Theoretical innovation occurred only when more advanced researchers challenged the scientific paradigm depicted in the textbooks.[10] Indeed, Kuhn, and more recently David Knight, depicted Boerhaave's *Elementa* as performing this function—transmitting the extant chemical paradigm of Robert Boyle.[11] Other historians have tried to situate Boerhaave's work as an instrument for transmitting "Newtonian" natural philosophy. This approach stems from Hélène Metzger, who cited Boerhaave's use of the "Newtonian" concept of attraction and the similarities between Newton's aether and Boerhaave's concept of fire.[12] Nevertheless, as Seymour Mauskopf has pointed out, Boerhaave's work has generally been lost in the historiography of eighteenth-century chemistry, which is dominated by the "canonical master narrative of the Chemical Revolution," centered on the theoretical interests and innovations of Antoine-Laurent Lavoisier.[13]

During the early eighteenth century, Boerhaave was revered for his teaching and for his orderly, encyclopedic presentation of medical subjects, including botany and chemistry. He lived in an age when the organization and presentation of knowledge was an important concern for philosophers and educators alike. Public anatomies, science courses, and textbooks became fashionable in the Netherlands and elsewhere, and as Lissa Roberts has argued, the methods deployed by university professors and public lecturers to organize their lectures and capture their audience's attention were often very similar.[14] Debates about the content, status, and relationship between bodies of knowledge were carried out in the various encyclopedias and dictionaries, which were always tied into programs for intellectual and social reform, as reformers attempted to reconfigure the intellectual landscape to support their own social and philosophical agendas.[15] In this respect, Boerhaave as a master organizer and pedagogue of medicine was at the forefront of the eighteenth-century drive for systematization.

This book rejects the view that Boerhaave's pedagogy of chemistry constituted the passive transmission of established knowledge. Rather, following revisionist work in the history of science, I present science pedagogy as an active, constructive endeavor that reconfigures scientific fields, opens new possibilities, and generates knowledge.[16] Boerhaave's *Elementa Chemiae* functioned equally as a pedagogical text for students and as an account of more advanced work addressing fundamental questions. The chemical novice could study the *Elementa* to gain a grounding in Boerhaave's chemical system, while the experienced natural philosopher could read the book for its novel discussions of chemical thermometry or burning lenses or Boerhaave's theory of menstrua (i.e., solvents).[17] This dual function would have surprised no one in the eighteenth century, when university professors, who were not required to publish (what we would call) "research," often communicated their discoveries and innovations in academic genres: textbooks and dissertations.[18]

This book examines the development of Boerhaave's chemical system as a product of thirty years of pedagogical process from his days as a student at Leiden through the publication of the *Elementa Chemiae*. Boerhaave used academic methods of structuring university medical courses and the experimental methods of demonstration from physics and physiology to reshape the pedagogy of chemistry and, in doing so, reform the status, aims, and theoretical framework of the field. He created his chemical system by defining problems, parameters, and methods; testing claims and theories; and ordering the whole according to the pedagogical methods that he and others used at the University of Leiden at the turn of the eighteenth century.

BOERHAAVE'S PEDAGOGICAL REFORM IN CONTEXT

In 1575 the curators and burgemeesters (i.e., trustees) of the newly founded University of Leiden commissioned Wilhelm Feugueraeus, a theology professor, to provide curriculum recommendations for the various faculties. In medicine he suggested that students perform "fewer disputations and speeches" but rather pursue "the inspection, dissection, dissolution, and transmutation of living bodies, plants, and metals."[19] These recommendations were based largely on the curriculum of the medical faculty at Padua, where during the 1530s and 1540s, as part of the medical humanist movement, professors had reformed the medieval curriculum and its emphasis on logical disputations. The proponents of this movement advocated a return to the ancient Greek "method" of medical practice, which the reformers characterized as undogmatic, empirical, and founded on the knowledge of *things*: medicinal herbs, specific diseases, and anatomical structures.[20] On the basis of the Paduan model, the Leiden curators established a botanical garden and an anatomy theater, with medical faculty professorships assigned to manage each facility and instruct students in each field.[21]

One might interpret Feugueraeus's recommendations for students to conduct "dissolutions" and "transmutations" as a call for chemical instruction, but this, however, would overstate the situation. In 1575 "chemistry" (*chemia*, or, just as often, *alchemia*) simply did not exist as a body of knowledge that could be taught at the university. Whereas botany (especially as materia medica) and anatomy had well-established traditions as medical fields, there was no tradition of chemistry per se within classical medicine. As the subject related to the reformed Hippocratic and Galenic medicine that Feugueraeus advocated, "chemistry" signified an ancillary and sometimes competing form of medical practice. A "chemist" was an artisan, usually an apothecary or herbalist, who made his living by selling medicaments, cosmetics, and other products of his workshop. Because these chemical practitioners worked with their hands and lacked academic credentials, they held a position inferior to that of university physicians in the medical hierarchy. (Many European cities, for example, commissioned boards of physicians to police medical practice, which included the activities of apothecaries.)[22] From the physician's perspective, the role of the apothecary was to fill prescriptions written by physicians, although most apothecaries also treated patients independently for minor illnesses. The business of chemistry as well as the training of new chemists were controlled by artisans' guilds. Novice chemists learned their trade by apprenticing themselves to a master chemist and working in his shop, not by studying the chemical arts from a book or by hearing lectures as university medical students did.

Despite these hurdles, by the first decades of the eighteenth century, chemistry had become an established part of the Leiden medical curriculum, institutionalized through a university-supported chemical laboratory and a chair of chemistry on the medical faculty. How did this happen? Among the many fundamental changes that shaped and reshaped chemistry as an art during the seventeenth century, none was more significant than the development of a pedagogical tradition. This didactic tradition began in 1597 with the publication of Andreas Libavius's *Alchemia*. As Owen Hannaway and, more recently, Bruce Moran have argued, Libavius literally "called chemistry into being" by extracting key terms, concepts, and operations from the various chemical arts and techniques and organizing them into a methodical series of definitions, precepts, and operations, which could be taught efficiently to novices. This act defined *chemia* as a body of knowledge distinct from its various applications and gave the new field an academic identity.[23] The chemical textbook tradition, which Libavius's *Alchemia* spawned, provided chemical practitioners with a forum for incorporating new knowledge into their field, evaluating and debating practical and philosophical issues, and presenting the chemical art to a wide audience.[24]

A corollary of Hannaway's thesis was that Libavius's didactic methods, committed to the clear and open presentation of knowledge, forced him to reject the secrecy and esotericism of the Paracelsian and alchemical traditions.[25] The implication of this claim was that the didactic textbook tradition was a major step toward the clarity and rationality of modern chemistry. This aspect of Hannaway's thesis, however, has generated serious criticism from other historians. Jole Shackelford warned that many Paracelsians were not as committed to secrecy as Hannaway claimed, and William Newman has pointed out that Libavius himself employed Paracelsian and alchemical tropes in his work.[26] Later didactic textbook authors freely incorporated Paracelsian and alchemical theories, terminology, symbolism, and processes into their work. Antonio Clericuzio and Bernard Joly have both argued that, in fact, textbooks functioned as one of the main vehicles for spreading Paracelsian ideas in the seventeenth century.[27] Newman and Lawrence Principe have likewise shown that textbook authors of the seventeenth century made no clear distinction between the terms "alchemy" and "chemistry" and the practices they signified, suggesting that historians employ the archaic term "chymistry" to identify this mixed form of the chemical art.[28]

In a recent essay, Lawrence Principe described an early eighteenth-century chemical "revolution nobody noticed" in which chemists at the Académie royale des sciences worked to remake their field by distancing their practices from those of "sooty empirics," alchemists, poisoners, and chemical mountebanks.[29] Like these chemists, Herman Boerhaave sought to make chemistry more acceptable within his institutional context, the University of Leiden. His contribution to this revolution was to remake the pedagogy of chemistry following academic standards and practices, and in doing so, he completed the work that Libavius began. Boerhaave transformed the didactic, textbook "chymistry"—an artisanal practice that focused on making *things* and used alchemical and Paracelsian constructs as its theoretical framework—into "chemistry"—a form of demonstrative science and experimental philosophy that focused on making theoretical *knowledge*, which could then be deployed in the making of things.

As its title indicates, this book is about how chemistry was "invented," that is, how it was created (or perhaps, distilled) out of the morass of techniques, philosophical approaches, and literary traditions that could at one time or another be called "chemical" (or "alchemical"). The term "invention" (*inventio*), as used here, derives from classical rhetoric and refers to the logical process (or dialectic) of finding arguments in debate. As reinterpreted under the influence of Ramist pedagogical theory, *inventio* became the process of determining what topical material belonged to subjects as scholarly disciplines.[30] From a formal, academic perspective, a university curriculum subject (such as medicine) was defined as an "art": a system of precepts exercised together toward some useful end. The "precepts" of an art were its general principles, which guided the practitioner as he applied the art to specific applications. As a professor created a new course, he first had to decide what the proper topics of discourse were for the "art" (invention), and then he defined and divided these topics into precepts and arranged them according to a "method" (*methodus*, from classical rhetoric, meaning a theory of ordering) for its pedagogical presentation.[31] The organized body of guiding precepts, now "methodized" into an ordered series of pedagogical topics (an *ordo*), constituted the art's theoretical knowledge or *scientia*.[32] This body of precepts comprised the content of the university course, that is, the theoretical principles of the art, and the rules and reasoning that determined how the principles were applied in practice.

When Boerhaave prepared his first chemistry course in 1702, he subjected the chemistry that he learned as a student and practiced as young physician in Leiden to this academic notion of "art." He did this in order to integrate his course more easily with the medical courses he was already teaching. Ultimately, Boerhaave hoped to elevate the status of chemistry at the university from a craft practice of ancillary importance to university medicine to an integral part of the medical curriculum. Through his pedagogical revisions, he fundamentally changed the aims, methodology, and theoretical structure of didactic chemistry as it had been previously taught at Leiden and elsewhere. Whereas previous chemistry courses at Leiden focused on recipes (for medicaments and other chemical products), Boerhaave's chemistry lecture course strove to examine the principles of nature and how they were applied in chemical operations. He presented these principles, following the method of his medical courses, as a logically ordered series of *theses*—short definitions, statements of fact, or arguments regarding the interpretation of a previously stated fact.

The heart of this book examines how Boerhaave "found the arguments" and precepts for his new, university chemistry, and how this process—the experimental testing of claims and operations—became a central part of chemistry. Throughout his academic career, when Boerhaave examined extant practices in chemistry, he inevitably found contradictory claims. A central, pedagogical problem for him was how one adjudicated these claims. He did this in several ways. The most direct way was by comparing claims with his own practices and beliefs, most of which derived from the extant practices and cultural climate of the Leiden university community. His yardstick was augmented by the work of authorities, such as Robert Boyle, whom he admired and trusted. Ultimately, however, he relied on testing chemical claims experimentally. This approach derived from extant chemical traditions: modes of alchemical experimentation, Robert Boyle's "sceptical" experimental narratives, and the tradition of chemical analysis of physiological fluids as practiced by the Leiden experimental physiologists. Boerhaave recounted many of these experiments in his lecture notes and in the *Elementa Chemiae*. The problem of evaluating chemical knowledge was so important to him that his later chemical courses (1718–29) and the *Elementa* presented his methods for testing claims as an integral part of chemical practice.

In the end, I argue that Boerhaave was one of the founders of what Arthur Donovan has called "philosophical chemistry."[33] Peter Shaw (1694–1763), chemist and translator of Boerhaave's *Elementa*, explained that the pursuit of philosophical chemistry would "by means of appropriate Experiments, scientifically explained, lead to the discovery of *Physical Axioms*, and *Rules of Practice*, for producing useful effects" in order to "improve the State of natural knowledge, and the Arts thereon depending."[34] Shaw's definition perfectly stated Boerhaave's program for chemistry. This shows the extent to which Boerhaave exerted an influence on Shaw and other practitioners of chemistry, enticing them to adopt his method and approach to the chemical arts. In Boerhaave's work they found a system for ordering chemical facts and operations and a methodology for evaluating and making chemical knowledge. His pedagogical reframing situated chemistry as both a form of natural philosophy and, also, a practical art to be valued for its utility. This characterization promoted chemistry's wider dissemination and acceptance, since it could easily be plugged into reformist arguments about the utility of the sciences.[35] Philosophical chemistry, with its emphasis on generating and evaluating *knowledge* in addition to *things*, was the root of the modern chemical sciences.

OVERVIEW

In examining Boerhaave's creation of his chemical system, I emphasize three main themes. First, his chemistry was the result of pedagogical process: solving problems of invention, ordering, and presentation that would allow him to integrate chemistry pedagogically and philosophically into the Leiden medical curriculum. Thus, the pedagogical work that he devoted to creating his chemistry courses and composing the *Elementa Chemiae* was a form of *doing* chemistry. Second, Boerhaave built his program for medicine and chemistry from resources available to him within the Leiden context.

He employed the material (library, laboratory, botanical garden) and cultural (pedagogical practices, standards of religious orthodoxy, commercial values) resources found within and around the University of Leiden community to fashion his system. Finally, he evaluated and incorporated parts of seventeenth-century "chymistry," including inferences he made from his own alchemical work, into his chemistry courses. Thus, Boerhaave's case supports the notion of a reinterpretation, rather than a radical break, between "alchemical" and "chemical" approaches to both theories of matter and the practical, chemical arts in the early modern period.[36]

This study also examines in detail several Boerhaave manuscripts for the first time. These manuscripts are currently held at the Fundamental'naya Biblioteka Voenno-Meditsinskoi Akademii (Fundamental Library of the Military-Medicine Academy, also called the Kirov Academy) in St. Petersburg, Russia. This archive contains most of Boerhaave's extant papers. Included in the manuscripts that I have examined are his autograph lecture notes from three different chemical courses, notes from various medical courses, chemical and medical reading notes, and laboratory notebooks detailing alchemical and other experiments.[37] Until now the *Elementa Chemiae* has been the main source of historians for their accounts of Boerhaave's chemistry. These manuscripts, however, detail courses from the whole of his university career. They have allowed me to document the development of his chemical pedagogy, so that I may present it as an ongoing process rather than as a static entity. His extant notebooks also allow for comparisons between his research interests and the presentation of that research in his courses and published works.

Historians of Dutch culture and science have remarked on the peculiar Dutch approach to art and the sciences: attention to empiricism, realism, and the detailed description of things.[38] Empiricism was an integral part of Boerhaave's natural philosophy too, but reference to the Dutch cultural milieu alone does not explain the development of his chemistry and medicine. I prefer to look at specific, local resources and practices found at the university and town of Leiden in order to uncover the roots of Boerhaave's system. Therefore, I examine local notions of religious and philosophical orthodoxy, the methods and theories of the Leiden professors, and the institutional resources of the university such as the library, physics theater, botanical garden, anatomy theater, and chemical laboratory. Chapter 1 discusses the development of Boerhaave's basic approach to medicine and chemistry by tracing his education as a student at Leiden and then his early program for curriculum reform as a lecturer on the Leiden medical faculty.

This chapter is framed (in part) in dialogue with work by Rina Knoeff, who has argued that Boerhaave's medicine was a reflection of a specifically Calvinist approach to nature, and the opposing view of Hal Cook, who argued that Boerhaave's Calvinism did not shape his medical system.[39] I argue that while Calvinism does not seem to have shaped the specific claims of Boerhaave's medicine, it did shape the direction of his medical study and, later, his pedagogy of medicine. Ultimately, I show that his medicine was eclectic, borrowing from diverse traditions, empirical, mechanical, and chemical. His real gift, and the key to his success, was his ability to portray his medicine as orthodox, that is, derived from the Hippocratic tradition and framed by Calvinist values.

Boerhaave shifted the pedagogical presentations of his chemistry courses away from the traditional didactic emphasis on chemical recipes and onto the theoretical precepts of chemistry. Chapter 2 addresses the origins of the didactic tradition in chemistry beginning with Libavius, its development through the seventeenth century, and the difficulties that its practitioners in Leiden had in integrating it into the university curriculum. This chapter challenges the Hannaway thesis, suggesting that didactic chemistry did not accomplish the revolution in chemistry that Hannaway saw in Libavius's writings. Rather, the chapter looks at what didactic chemistry became during the century and examines how and why Boerhaave rejected many of the goals and theoretical constructions (i.e., chemical principles) of that tradition. Instead, as a private physician in Leiden, he began a program of testing chemical claims, which shaped his later approach to teaching chemistry. His skepticism regarding the theoretical framework of didactic chemistry inspired him to devise a new pedagogical method and theoretical framework for his first chemistry course in 1702. Chapter 3 examines this reframing and discusses how this theory related to both academic pedagogy and the didactic tradition. I analyze the origin and structure of his lecture course, focusing on the development of his instrument theory, which acted as both a theoretical framework and pedagogical structure for his courses.

One way in which chemistry became established as a legitimate academic discipline was through its association with established medical fields, like materia medica, physiology, and pathology. Chapter 4 addresses the various ways in which Boerhaave integrated chemistry into his medicine and examines the ramifications for both fields. I look at how he defined the place of chemistry in the Leiden medical curriculum, and also how he used chemical analysis, concepts, and terminology to understand human physiology and pathology in his medical courses. I argue, in opposition to some historians' characterization of Boerhaave's medicine as "mechanistic," that he was also, if not primarily, a chemical physician and that his medical work presented a relatively sophisticated form of chemical medicine. As such, chemical knowledge was necessary for any student or physician who wished to understand Boerhaave's medical system.

Boerhaave's pedagogical reforms in chemistry came to fruition in his instruments course, which he presented over ten years, from 1718 to 1728. In these lectures, he opted to convey the theoretical principles of his chemistry through demonstration experiments performed in the Leiden chemical laboratory. Chapter 5 examines Boerhaave's reasons for adopting this new style of presentation, which aimed not only at showing theoretical principles in action, but also at explicating the experimental methodology through which those principles were established. I also investigate the experiments that Boerhaave undertook to work out his demonstrations. This work led him to novel approaches and conclusions, especially in chemical thermometry and in the chemistry of air. Ultimately, I show that his pedagogical work in assembling the instruments course was also a form of experimental natural philosophy.

Most previous accounts of Boerhaave's chemistry have focused on his monumental textbook, *Elementa Chemiae* (1732). This work was the most mature statement of Boerhaave's chemistry and was the compilation of thirty years of pedagogical process. Chapter 6 presents the *Elementa* as both a synthetic and expanded version of all of Boerhaave's previous chemistry courses. He structured the book according to his previous pedagogical methods, but he also realized that his system was incomplete. He pointed out gaps in the chemical knowledge he presented and suggested strategies for further research. In this regard, the *Elementa* contained much discussion of current questions in chemistry, how to test chemical claims, and how to establish new, chemical facts. Thus, the critical and experimental methodologies that Boerhaave employed to create his chemical system were an integral part of the knowledge content of his textbook and made the *Elementa* the first textbook of "philosophical chemistry." I conclude the chapter with an examination of Boerhaave's positions on three topics that historians have identified as central to the development of chemistry in the eighteenth century—the chemistry of air, the phlogiston theory, and the notion of chemical affinity.

Chapter 7 investigates Boerhaave's experimental work in transmutational alchemy and the composition of metals. This work provides an opportunity to examine how Boerhaave applied his pedagogical method in nonpedagogical contexts in order to structure experimental data. In effect, he used his pedagogical method to present experimental evidence and make logical arguments based on this evidence. His alchemical project spanned his entire career, from the 1690s through the publication of three papers on the chemistry of mercury in the *Philosophical Transactions of the Royal Society of London* (1733–36). The chapter focuses on Boerhaave's experimental work and the preparation and writing of these papers on mercury. Specifically, he tested claims associated with the mercurialist school of alchemy—those who argued that mercury alone was the basis of all metals and the route to the philosophers' stone. In the end, Boerhaave used his method to demonstrate that the central claims of the mercurialist school of alchemy were false.

Like the Enlightenment reformers who came after him, Boerhaave saw the reform of chemistry as a process—one that he did not complete but for which he strove to lay the groundwork. Boerhaave's success in this endeavor cannot be measured by the theoretical constructs associated with his name (of which there are few), but rather by the drastic redefinition of chemistry that he accomplished directly through his pedagogical work. After the publication of the *Elementa Chemiae* (1732), chemistry emerged as a viable and accepted part of the university curriculum, spawned in part by the efforts of his students to replicate his method at other institutions. The pedagogical structure and theoretical framework of the *Elementa* became, in effect, the structure and framework for academic chemistry during the eighteenth century. Most chemical courses, even non-university ones, such as Guillaume-François Rouelle's in Paris and Peter Shaw's in London, adopted some form of the instrument theory and advocated Boerhaave's experimental methodology for establishing chemical facts and "axioms." Boerhaave, more than any other chemist, was the architect of this new chemistry.

CHAPTER ONE

Medicine as a Calling

Eighteenth-century biographical treatments of Herman Boerhaave universally praised his intellectual qualities: his encyclopedic knowledge of medicine and the sciences, his mastery of classical and modern European languages, and his acute memory. Almost all of these accounts also posit that his greatest attribute was his strength of character. Take, for example, Samuel Johnson's description of Boerhaave in the *Gentlemen's Magazine* in 1739: "But his knowledge, however uncommon, holds, in his character, but the second place; his virtue was yet more uncommon than his learning. He was an admirable example of temperance, fortitude, humility, and devotion."[1] According to his students, Boerhaave's virtues manifested themselves in his daily life and work. Albrecht von Haller, who studied with Boerhaave in the mid-1720s, stated that despite his wealth at the time, "Boerhaave lived like a poor brewer." He was amazed at Boerhaave's capacity for work, filling the entire day from seven in the morning until dusk with teaching, tending patients, and conducting correspondence.[2] Boerhaave's student and biographer William Burton concurred with Von Haller's assessment. Regarding Boerhaave's appearance, he stated: "He was negligent of dress, and in his gate and deportment there was an honest and somewhat awkward simplicity." Yet Burton described Boerhaave's habits as "temperate in every thing except application," explaining that he rose at four o'clock in the morning during summer and five o'clock in winter, whence he removed himself to his unheated study, first for an hour of prayer and then for a few hours of reading.[3]

To eighteenth-century readers, these examples of Boerhaave's restraint,

FIGURE 1. Portrait sketch of Herman Boerhaave. Courtesy of Fisher Collection, Chemical Heritage Foundation Collections. Photograph by Gregory Tobias.

modesty, and industriousness were understood within a religious context that has often been missed by modern readers. As Max Weber argued almost a century ago, in Protestant societies where Catholic doctrines regarding the intentionality of sin were rejected, proper *behavior* was the only acceptable sign of one's religious devotion and membership in the elect. The restraint of passion, emotion, and flamboyance, coupled with a pious zeal for one's work (i.e., one's "calling"), was a sign of divine grace.[4] Boerhaave, steeped in the tradition of Dutch Calvinism, was a true believer in the Reformed faith, and he labored to make his public conduct fit this ideal model of Calvinist piety. In his funeral oration for Boerhaave, Albert Schultens, a theologian at Leiden, related that he once asked Boerhaave how he was able to maintain his patience, even under great provocation. Boerhaave replied that "he was naturally quick of resentment, but that he had, by daily prayer and meditation, at length attained mastery over himself."[5] Andrew Cunningham has characterized Boerhaave's conduct as deriving from an "eirenic" (i.e., peace-loving) stance similar to the position of Robert Boyle and the Latitudinarians in England. According to Cunningham, adherents of the "eirenic" stance rejected sectarian dogma and metaphysics in both religion and philosophy, promoted toleration, and held that "the individual should strive for a certain inner calm or imperturbability of spirit, and express his love of God through applying himself wholeheartedly to *work*."[6] The work that Boerhaave ultimately undertook to express his devotion was medicine and the teaching of medicine. He saw medicine as his calling.

Several scholars besides Cunningham have investigated the impact that Boerhaave's Calvinist faith had on his approach to medicine and chemistry. At issue was the origin of Boerhaave's empirical philosophy, upon which he constructed a medical system that privileged sensory experience (through anatomical and clinical observation and experiment) over theoretical approaches of any kind. Historians who have studied Boerhaave have observed that, although he had a reputation for piety, the content of his work did not appear to be shaped overtly by religious ideology. On this basis, Harold Cook has argued that Boerhaave's empirical stance derived from debates among philosophers in the seventeenth-century Netherlands regarding the importance of quelling the passions in order to free one's mind from bias and produce reliable knowledge. As he asserted, Boerhaave maintained that the simple use of will or reason was often insufficient to quell the passions, and thus medical systems based on reasoning from theoretical assumptions were fallible. In light of this realization, Boerhaave turned toward philosophical empiricism.[7] Rina Knoeff, however, has argued that Boerhaave's empiricism derived directly from his Calvinist faith. She suggests that the fallibility of the human mind was a tenet of orthodox Calvinism; God as well as the ultimate causes that shaped the natural world were fundamentally unknowable. Nevertheless, Calvinists were called to study the natural world as a means of praising God and his creation. From this perspective, Knoeff argued that Boerhaave's medicine was a form of natural theology, but one in which the structure of the world was merely described without theoretical speculation about final causes.[8]

These two views are not necessarily incompatible. As Gerrit Lindeboom pointed out, Boerhaave was notoriously opaque about his private life and thoughts,[9] so it would be nearly impossible to decide the question of the true origins of his medical philosophy. As an alternative view, I suggest that these two perspectives, the "Calvinist" and the "philosophical," constituted mutually supporting sources for Boerhaave's empirical approach to medicine, chemistry, and natural philosophy. He, in fact, did not see these two approaches as separate. As a committed empiricist, he advocated for the use of methodological tools to help the physician discipline his perceptions and his use of reason. These included the precept that general claims in medicine must be based on repeated and repeatable empirical observations, a methodology derived from the philosophical work of Francis Bacon.[10] Complex phenomena were to be modeled though the application of mathematics, a method that mirrored the approach of the practitioners of "physico-mathematics," such as Isaac Newton and Christian Huygens.[11] Boerhaave asserted, however, that these methods were only viable in the hands of a capable practitioner, by which he meant a physician who possessed the same self-discipline, devotion to community, and commitment to work that the ideal Calvinist did. Only by shaping one's character and behavior in accordance with these values could a physician make reliable choices, whether it be determining a program of treatment for a patient or creating and evaluating new medical knowledge. For Boerhaave, these two elements, the empirical and mathematical philosophy and "Calvinist" self-restraint and ethics, constituted the same methodological approach to medicine. He called this approach to medicine "Hippocratic."

Within the University of Leiden community, Boerhaave's reputation as both a model Calvinist and a Hippocratic physician served an important rhetorical function during his medical career. His perceived orthodoxy and dedication to extant medical practices gave him credibility within the Dutch cultural context of the early eighteenth century. Some historians, in fact, have argued that Boerhaave's medicine was simply a reflection of traditional practices in Leiden, and that he was merely in the proper time and place to be able to transmit the indigenous, Dutch medicine to a wider audience.[12] Certainly, much of his medicine was shaped by the ideas and practices entrenched in the philosophy and medical faculties of Leiden. Yet it would be inaccurate to say that he merely passed on the traditional medicine. In terms of its content and methodology, Boerhaave's Hippocratic medicine was thoroughly modern and, in many ways, novel. His achievement was his ability to present successfully his new medicine as derived from established practices and suffused with traditional values and thus gain the support of traditionalists at the university. In effect, he generated support for his medical reforms (which included his new chemistry) by associating his Hippocratic medicine with the values and character traits that he himself embodied.

EDUCATION

All depictions of Boerhaave's education derive from his "Commentariolus," an autobiographical collection of notes that he composed near the end of his life. In this text, Boerhaave asserts that his education formed his mind and character into a model of Calvinist piety, industriousness, and mental discipline. He describes how, although he was not born with a tireless work ethic and disciplined mind, he fashioned himself into an exceptional student, an orthodox Calvinist and, finally, a Hippocratic physician.[13] The aim of the "Commentariolus" and other autobiographical references in his public orations was to reinforce Boerhaave's public image as a pious man of good character and industry, whose behavior and opinions were shaped through a "natural" and just point of view, unfettered by sectarian bias. This interpretation has shaped most subsequent biographical accounts of Boerhaave. William Burton (1703–1753), a former student, collected the notes that comprise the "Commentariolus" after Boerhaave's death and used them in part to compose the first biography of Boerhaave. That biography and Albert Schultens's funeral oration for Boerhaave, which was also based on the "Commentariolus," became the two main published sources of information about his life during the eighteenth century.[14] Thus, Boerhaave played a central role in shaping the public accounts of his own character and its development. The account of his intellectual development presented here also uses the "Commentariolus" and other public pronouncements by him, but with the understanding that they are rational reconstructions designed to promote a specific view of Boerhaave's life and work for specific rhetorical purposes.

According to Boerhaave's own account, the process of shaping his character began in his youth. He was born on New Year's Eve, 1668, in the village of Voorhout, just outside of Leiden, to Jacob Boerhaave (1625–1683), a *predikant* (preacher) in the Dutch Reformed Church.[15] Jacob intended for Herman, as his eldest son—he had previously buried five children and one wife—to follow in his footsteps and become a minister. In the "Commentariolus" Boerhaave describes his father in the following manner: "He was an open, honest, and simple man: a head of the family of the greatest love, care, diligence, frugality, and prudence. Though not rich in material things, but full of virtue, he presented to his nine children a singular example of what may be accomplished through strict thrift and frugality."[16] Comparing Boerhaave's description of his father with his biographers' descriptions of himself, one can only conclude that he saw his father as the model for his own conduct. In contrast with contemporary accounts that remark on the permissiveness of Dutch parents toward their children, Jacob took an active role in instilling in his eldest son the discipline and industriousness proper for a Calvinist minister. Jacob encouraged in him the love of study, but he also taught young Herman the value of hard work. As he recounted in an oration fifty years later, "With Socratic care he [Jacob] cultivated in me from my earliest youth onwards a love for the study of the humanities; at the same time, I was physically hardened through being made to work in the fields, and he saw to it that I would not grow lazy or languid."[17] There was a practical value in Jacob's character-building lessons: supporting (eventually) nine children on a clergyman's meager salary was not an easy task, and everyone would have had to do his part to help make ends meet. But Boerhaave's labor also reflected a central value in Dutch Protestant culture: that work was the cure for vice and instilled in the working individual habits that were productive and useful to society as a whole.[18]

The Reverend Boerhaave oversaw his son's early education in the classics in order to prepare him for university studies and the ministry. In his biographical notes Herman Boerhaave described his father as a classically educated man, and Jacob envisioned a similar type of education for his son. Young Herman studied Latin and Greek grammar, the classics, the Bible, and Christian Matthias's *Theatrum Historicum Theoretico-Practicum* (first edition, 1648), a survey of "universal" history composed by the chair of history at Leiden. Boerhaave reported that by the age of eleven he could read Latin and Greek easily and was able to translate the classical languages into Dutch and vice versa. In July 1682, at the age of thirteen, Jacob sent the young Herman to the Latin school affiliated with the University of Leiden as a preparation for his university studies in divinity. As the name indicates, Latin schools in the Dutch Republic primarily taught Latin grammar and literature, but after the student had progressed sufficiently, logic, rhetoric, and a smattering of Greek were also available.[19] After examining the boy, the headmaster, Wigard Wijnschoten, placed Boerhaave in the fourth form, indicating that he had already mastered the rudiments of Latin grammar and literature. By February 1684, he completed the sixth and highest form, at which point he was permitted to matriculate into the liberal arts (or philosophy) faculty of the university. He delayed his matriculation, however, and remained at the Latin school for an additional semester. As he later wrote, he was recovering from a large ulcer that had developed on his leg, which he cured himself with a concoction of salt and his own urine.[20] What he did not record was that he was also in the midst of a family tragedy. In November 1683 Jacob Boerhaave had unexpectedly passed away.

Jacob's death placed a financial burden on Herman, which had to be solved before he could begin his university studies in earnest. In accordance with Dutch law, Herman, as the eldest son, was eligible to inherit a substantial portion of his father's assets and property.[21] The sum that he received was probably not great, and much of that was devoted to supporting his family—his stepmother (Herman's mother had died in 1673) and eight siblings. Still intent upon entering the ministry, Boerhaave decided to use a portion of his inheritance to begin his university education, and he matriculated into the arts faculty of Leiden in August 1684. Jacob Trigland (the younger, 1652–1705), a friend of Boerhaave's father and professor of theology, acted as an adviser to Boerhaave and assisted the young student in obtaining a scholarship from the curators and burgemeesters of the university. In December 1687 Boerhaave received one of thirty theological scholarships to the States College (called the Collegium Theologiae in the university registry), where students lived under disciplined conditions, attended theology lectures at the university, and eventually took the clerical examinations administered by the Reformed Church. Boerhaave, however, in an unusual situation, was allowed to live in his stepmother's house, and he was also awarded nine months' worth of his stipend up front.[22] At this time, he matriculated into the theology faculty and had just over five years of financial support for completing his studies.

On the advice of Trigland, Boerhaave studied in the liberal arts faculty under the tutelage of Wolfred Senguerd (1646–1724). For a prospective theology student and Calvinist minister, Senguerd was the safest and best choice because of his doctrinally sound teachings. He had been appointed professor of philosophy in 1676 in the wake of the controversy over Cartesian philosophy, which had caused dissension in the university for three decades.[23] Senguerd was seen as uncontroversial and somewhat conservative. The university curators, in fact, presented him to the university community as an Aristotelian and promoted his appointment as a calming influence: "for a better maintenance of good order."[24] In reality, Senguerd was a philosophical eclectic, but he avoided controversy and took pains to frame his philosophy within the parameters of orthodoxy. In his textbook *Philosophia Naturalis* (1680, 1685), he adopted the topical structure and terminology of Aristotelian physics (causes, forms and matter, motion), but within this traditional structure he infused modern ideas. He replaced the Aristotelian plenum with a world of corpuscles and void that a reviewer in the *Acta Eruditorum* called "half-way between the Cartesians and modern Democritans." In practice, Senguerd presented a multitude of opinions in his lectures on natural philosophy, but guided his students toward those that would not generate theological or philosophical controversy. For example, he discussed in detail the cosmological systems of Copernicus, Ptolemy, and the "Cartesian system," but he promoted the system of Tycho Brahe, because he felt that it accounted for all of the observed phenomena but did not violate the theological tenet of the immobility and centrality of the earth. Similarly, when discussing the phenomenon of gravity, he explained and rejected both Descartes's and Aristotle's positions, arguing that God was the primary efficient cause of motion. This view included the motion induced by gravity, which he stated was a type of "impetus" implanted in matter by God and perpetually conserved in bodies.[25] In this way, Senguerd presented his students with a broad survey of philosophical approaches, but he advocated only those within the accepted bounds of orthodoxy.

Under Senguerd, Boerhaave became one of the star students of the Leiden arts faculty. As part of the academic culture of the arts faculty, Boerhaave participated in five academic disputations during 1687–88 conducted in the grand auditorium of the main university building. During each of these exercises, he was required to defend in public debate a number of previously circulated philosophical theses usually composed by the student's advising professor, in this case, Senguerd. As such, the theses reflected Senguerd's interests in physics and metaphysics: one set of theses examined the cohesion of bodies ("De Cohaesione Corporum") and three concerned the faculties and powers of the mind ("De Mente Humana"). Since the university statues only required arts students to participate in one disputation for their degree, Boerhaave's disputations represented Senguerd's attempt to use a gifted student as an ideological mouthpiece to assert his and the arts faculty's philosophical orthodoxy: each set of theses criticized and rejected the Cartesian position.[26]

Boerhaave's status as a top student and ideological mouthpiece for the arts faculty was reinforced in 1689 when the professors of that faculty selected him to petition the *rector magnificus* (i.e., chancellor) for permission to give an oration to the university body. The curators and burgemeesters had established this practice two years earlier as a means of rewarding promising students. Those who delivered such an oration were awarded a gold medal on which the student's name and the title of the oration were inscribed.[27] Boerhaave's oration concerned the obtuse topic of Cicero's interpretation of Epicurus's doctrine of the greatest good. Although ostensibly an examination of ancient views on ethics, the oration, in effect, constituted a refutation of Pierre Gassendi's attempt to Christianize Epicurius's ethical theories and, by extension, his physical theories. Boerhaave composed the oration under the direction of Jacob Gronovius (1645–1715), a professor of classical languages, and, like his earlier disputations, this exercise was designed to demonstrate the arts faculty's support of Calvinist orthodoxy. Boerhaave deployed classical authors to argue that Epicurus was a sensualist and a libertine, whose depravity stemmed from his denial of a spiritual nature and, ultimately, of God. By contrast, Boerhaave maintained that self-restraint was a necessary philosophical and religious virtue whose seat was the human soul.[28] The implication of Boerhaave's thesis was that philosophical truth stemmed from theological orthodoxy, a position that was proper for an aspiring minister and that supported the prevailing, public ideology of the university.

While Boerhaave was fulfilling his role as top arts student, his interests seem to have shifted from the conservative Senguerd to the progressive and mathematical camp of Burchard de Volder (1643–1709).[29] Unlike Senguerd, De Volder was at the center of the Cartesian controversy in Leiden, and even after the censure of Cartesian methods in 1676, he still espoused a modified form of this philosophy. He was a Mennonite who was appointed to arts faculty in 1670 to teach logic and then physics, but only after he had converted to the Reformed Church. His extant course notes and disputations show that the foundation of his ideas regarding matter, motion, and cosmology clearly derived from the philosophical work of René Descartes.[30] As Cartesian methods in philosophy became increasingly marginalized at the university, De Volder explored new approaches. Inspired by a trip to the Royal Society of London in 1674, where he witnessed experimental demonstrations performed during meetings of that body, he inaugurated the teaching of *physica experimentalis* in Leiden. De Volder's *theatrum physicum* was the first such experimental demonstration space in any European university. Here, he presented to students, once a week, demonstration experiments on pneumatics, optics, thermometry, pendulums, devices to show centripetal force, and the dissection of frogs. The centerpiece of De Volder's demonstrations in the *theatrum*, however, was an air pump of his own design, built by the famed Leiden instrument-maker Samuel van Musschenbroek (1639–1681).[31]

While he was devising experimental methods of pedagogy, De Volder also began to cultivate his interest in mathematics, especially ideas regarding the use of mathematical methods in physics or "physico-mathematics." He was appointed to the chair of mathematics in 1681, and during the next decade he increasingly began to see mathematics as the science of pure reason and the road to certainty in philosophy. For De Volder, mathematics described the general properties of bodies in a clear and precise manner, and since mathematical operations only proceed from necessary consequences, one could deduce necessary (and therefore, certain) consequences from them. Such methodology applied to natural philosophy, he argued, would reveal previous errors, discipline the human intellect, and banish the "hallucinations" to which the unaided mind was prone.[32]

De Volder's approach to mathematics appealed to Boerhaave. He began the study of geometry, trigonometry, and algebra under De Volder in 1687, right about the time that De Volder was working through Isaac Newton's *Principia*.[33] Like De Volder, Boerhaave appreciated the power of mathematics to discipline the mind and bring certainty to claims in natural philosophy. As he explained in the "Commentariolus," these studies were "wonderfully pleasing to his disposition," and he admired "most of all the geometrical synthesis of the ancients, cultivating it in order to increase his intellectual power, & [cultivating] the analysis of more recent writers in order to discover new uses."[34] When Boerhaave graduated from the arts faculty in 1690, De Volder was his "promoter" (the opponent in the ceremonial disputation over the student's thesis, usually the student's adviser). As part of his graduation requirements, he composed a dissertation to defend in debate during his graduation ceremony. His dissertation, "De Distinctione Mentis a Corpore," revealed the influence of De Volder. In it Boerhaave argued that the body and mind (or soul) were distinct and separate entities, a thesis that had the strength of being both orthodox and amenable to Cartesian arguments. He posited that whereas the body had extension, may be divided into parts, and may be given motion, the mind was perfectly simple and could not be divided or have motion. Thus, employing essentially Cartesian definitions of "body" and "mind," he upheld the Calvinist doctrine of the soul against such notable materialist-atheists (in Dutch Calvinist opinion) as Epicurus, Thomas Hobbes, and Baruch Spinoza.[35]

In his "Commentariolus" Boerhaave suggested that he was largely self-taught in theology and medicine. Indeed, when in 1690 he began to study both fields in earnest, he was greatly assisted in these endeavors by obtaining a position at the Leiden university library. The library had, in fact, just acquired the famed book collection of Gerard and Isaac Vossius, which greatly expanded the library's holdings in recent theological and scientific texts.[36] Over the next three years, Boerhaave seemed to have read the library's holdings in both theology and all the medical fields, including anatomy, physiology, and chemistry. The claim that he was an autodidact originated from Boerhaave himself and was reiterated by all of his important eighteenth-century biographers.[37] He never matriculated into the Leiden medical faculty, and there was little evidence of him attending medical courses regularly. Nevertheless, to characterize Boerhaave as a complete autodidact would be incorrect. In the "Commentariolus," he observed that in theology he studied the Illustrationes Antiquitatum Hebraicarum (Illustrations of Hebrew Antiquity) with Trigland and ecclesiastical history with Frederick Spanheim (who was also the librarian), under whom he also participated in an academic disputation. Even in medicine Boerhaave had mentors, whom he attempted to emulate during his medical career. He attended the public anatomical dissections conducted by Anton Nuck (1650–1692), the professor of anatomy. These dissections, which took place during the winter when a cadaver was available, were open to all university students. He also developed a personal relationship with Nuck and visited the professor's private laboratory at his house, where Nuck had laid out large portions of the human lymphatic system. He had extracted the vessels from cadavers and preserved them by injecting them with a solution of mercury, lead, and pewter that he had devised himself. He also performed dissection and vivisection experiments on dogs, primarily to demonstrate the circulation of the blood and various surgical techniques. Boerhaave stated that he undertook similar experiments himself.[38] In addition, he attended the medical lectures of Nuck's colleague, Charles Drélincourt (1633–1697), but because he stated in his "Commentariolus" that he did this "shortly before his [Drélincourt's] death" (in 1697), these lectures may have taken place after Boerhaave's student days.[39]

The results of Boerhaave's self-study brought him remarkably in line with the orthodox positions of the Leiden theology and medical faculties. He stated that when be began to study theology as a student, he read the writings of the church fathers, beginning with Clement of Alexandria and progressing, in chronological order, to more recent authors.[40] In the "Commentariolis," he asserted that he "revered the honorable simplicity of their pure doctrine, the sanctity of their discipline, and the completeness of a life devoted to God." By contrast, "he grieved at the subtlety of the schools that had soon defiled theology." Boerhaave complained that the metaphysical speculations of Plato, Aristotle, Thomas Aquinas, Duns Scotus, and even Descartes were being used to amend scripture, with each sect claiming that its own metaphysics revealed the divine mind. Such behavior only provoked disputes, hatred, and discord among Christians and undermined the evident simplicity of the doctrines contained in scripture.[41] As Andrew Cunningham has astutely pointed out, Boerhaave privileged the writings of the older authorities—the church fathers—as being the most reliable expositors of the true interpretation of scripture. In effect, the older the author, the closer he was to the source, and the more reliable his doctrine.[42] Subsequent authors, however, had corrupted the older, pristine theology through the misapplication of philosophy, specifically speculative metaphysics. This approach to religious doctrine was firmly engrained in the Dutch Calvinist tradition, especially among the conservative theologians at the university, exemplified by Boerhaave's two teachers, Spanheim and Trigland, who wished to refute the doctrines of the "papist" scholastics and, more recently, the Cartesian-influenced Cocceians.[43] Although Boerhaave described his arrival at this theological position as the result of independent study and contemplation, he ended at the received view of his theological teachers.

When Boerhaave began to study medicine, he applied the same method that he used to study theology and arrived, similarly, at the view of his medical mentors. As he recorded in the "Commentariolus," he began with a careful reading and analysis of the Hippocratic Corpus and progressed chronologically through the Greek and Latin authors until he realized that "the later writers owe everything good [in their work] to Hippocrates." He then skipped to more recent authors, spending extra time to "contemplate" (*versat*) Vesalius, Fallopius, and Bartholin, proceeding until he had reached the "second Hippocrates," Thomas Sydenham, known for his empirical studies of epidemic diseases.[44] As Cunningham observed, the parallel between Boerhaave's theology and medical studies was striking.[45] Whereas Boerhaave praised the earlier interpretations of scripture—the church fathers—and found the true theology in their writings, he also esteemed the earliest formulations of medicine—the Hippocratic Corpus—and found proper methods for medical practice there. He contended that just as metaphysical speculation had corrupted the true interpretation of scripture, metaphysical speculation had corrupted the true medicine. Like religion, medicine became divided into "sects."[46] As the source of the true medicine, Boerhaave's Hippocrates was a strict empiricist who rejected metaphysical speculation and followed a method of collecting observations and remedies, in effect building a history of ailments and their cures. This empirical philosophy was, in fact, nearly identical to the position of Boerhaave's medical mentors, Nuck and Drélincourt. Similarly, all of the modern authors whom Boerhaave included in his studies and later recommended to his students were anatomists or experimentalists whose work he could interpret within this framework.

For economic and practical reasons, Boerhaave traveled about sixty miles north of Leiden to the Guelders Academy in the town of Harderwijk to receive his medical degree. The cost of graduation at the Guelders Academy was about half of the cost of a degree at Leiden, and one could obtain this degree quickly. The single professor who comprised the medical faculty in Harderwijk, Theodorus van de Graeff (d. 1701), was only too happy to augment his modest salary with graduation fees from short-term students. Boerhaave arrived in Harderwijk on July 12, 1693, with his thesis, "De Utilitate Explorandorum in Aegris Excrementorum ut Signorum," in hand. He matriculated into the academy and was examined by Van de Graeff over the next two days. On July 15, with Van de Graeff acting as his "promoter," Boerhaave defended his thesis during a small ceremony in the presence of the rector.[47]

Officially, Boerhaave was still a theology student at Leiden, but soon after his return from Harderwijk he was forced to abandon his plans for the ministry. As recounted by William Burton, he was traveling on a canal barge one day when he overheard a group of passengers condemning the philosophical doctrines of Spinoza. When one passenger "opposed . . . this philosopher's pretended mathematical demonstrations [with] only the loud invectives of a blind zeal," Boerhaave calmly interjected, asking the individual if he had ever inspected any of Spinoza's works. This remark silenced the discussion, but soon afterward rumors began to circulate that Boerhaave had become a Spinozist.[48] Baruch (or Benedict) Spinoza (1632–1677) was the Netherlands' most reviled author. The Reformed Church condemned his writings as blasphemous and atheistic, and the States of Holland (Holland's governing body) banned his books twice.[49] Boerhaave had argued against Spinoza in the thesis he composed for his arts degree, but he learned, probably from his teacher Trigland, that even a groundless rumor of sympathy toward Spinoza was enough to tarnish his name in the mind of the Reformed Church and block his ordination into the ministry. Thus, Boerhaave discontinued his theological studies, even though, as he stated in the "Commentariolus," he was already preparing a sermon to deliver after his ordination ceremony. Ironically, the topic of his oration was why, in previous times, the unlearned made so many converts to Christianity but that nowadays the most learned made so few.[50]

The canal boat incident was an immensely important and revealing episode in Boerhaave's life. In the "Commentariolus" he does not give any specific details regarding the incident, stating only that he was "inadvertently implicated in an event, although he was innocent."[51] What the "Commentariolus" and all eighteenth-century accounts of Boerhaave omitted was that he had probably read Spinoza at the time of the canal boat incident. An inventory of his books made after his death included Spinoza's banned *Tractatus Theologico-Politicus* (1670), and in a letter dated 1710 he recommended passages from the *Tractatus* to a former student.[52] The route through which Boerhaave was introduced to Spinoza was probably Burchard de Volder, who was well acquainted with Spinoza's works and may have met Spinoza in Amsterdam.[53] Thus, Boerhaave had probably read some of the *Tractatus* or at least discussed Spinoza's ideas with De Volder by the time of the canal boat incident (ca. 1693). As Rina Knoeff has pointed out, Boerhaave was not a dedicated follower of Spinoza; they disagreed on several fundamental issues, notably on their understanding of the nature of God.[54] Yet, this mistake in Boerhaave's public conduct revealed that he had a much broader and more liberal intellectual life in private than his public persona would suggest. In private he took great pride in his self-study and developed a confidence in his own judgment on philosophical matters. He would never pass judgment on a text without first examining it himself, regardless of what it was. Even the most reviled author could have things of value to say for the orthodox physician or Christian. As Boerhaave told his student, in Spinoza's *Tractatus* he appreciated the application of mathematical reasoning to problems and also Spinoza's doctrine of intellectual freedom.[55]

Boerhaave's fall at the hands of public rumor did not sway the support of his patrons. After he abandoned his theological studies, Johannes van den Berg (1664–1755), the secretary of the Leiden curators who had followed Boerhaave's career at the university, assisted the young physician in establishing a medical practice in Leiden. Still living in the house of his stepmother, he now supported her and his siblings by visiting patients and tutoring university students in mathematics, a practice he had begun while he was still a student.[56] Soon, he was invited by a courtier of Stadholder-King William to come to court at the Hague. He refused the invitation, however, stating in the "Commentariolus": "He [i.e., Boerhaave] was clearly content with a free life removed from the crowds and singularly devoted to adorning his studies further, where he would not compelled by others to say and feign things, to think but conceal others; rather, he would be seized and guided by studies of his interests." His obligation to his family and his disdain for court life in favor of a peaceful life of contemplation and study kept him in Leiden. In his free time, as before, he studied medical texts, performed chemical experiments in a home laboratory, and read the Bible and "those authors who profess to teach a sure manner of loving God."[57]

PATH TO PROFESSORSHIP

Boerhaave became a medical lecturer at the University of Leiden in 1701 during an acute period of hardship for the medical faculty. For the preceding ten years, the university had experienced a financial crisis, caused in part by the debt it incurred from the purchase of the Vossian library.[58] One result of this crisis was that the university had a great deal of difficulty in attracting prominent physicians to the medical faculty. During the decade from 1687 to 1697, each of the five chairs on the Leiden medical faculty fell vacant at least once. The university curators attempted to find "famous" men to fill these vacancies, but they found that they could not offer sufficient salaries to attract suitable candidates. The Dutch stadholder, William of Orange (who was also King William III of England), exacerbated the shortage of medical professors by blocking the appointment of professors who did not meet the standards of his orthodox, Calvinist supporters. These included anyone who was seen as a supporter of "Cartesian" philosophy, such as the "prefect" of Leiden's chemical laboratory, Jacob le Mort (1650–1718), and the Leiden graduate and medical lecturer Johannes Broen (d. 1703).[59] By the middle of 1701, the Leiden medical faculty had only two professors in residence, Peter Hotton (1648–1709), the professor of botany, and Frederik Dekkers (1644–1720), the professor of *praxis medica*. Many core medical courses were being taught by short-term lecturers or were simply not being taught.[60] As a result of this situation, matriculations and degrees granted in the medical faculty began to drop sharply after about 1690.[61]

Boerhaave and his patron, Johannes van den Berg, took advantage of this situation to have Boerhaave appointed a medical lecturer, even though he did not seem a likely choice at the time. In 1701, Van den Berg (secretary of the curators and now president of the Leiden burgemeesters) suggested, following recent precedent, that Boerhaave be hired as a temporary lecturer to teach some of the neglected courses in the medical faculty. At the May 18 meeting of the curators and burgemeesters, he was appointed as a lecturer for the *institutiones medicae* course for a period of three years at 400 guilders per year.[62] Boerhaave's lectures were a success, and in January 1702 the curators and burgemeesters granted him permission to conduct "private" lectures—meaning that he would collect lecturing fees directly from students—in anatomy and chemistry, as the resolution stated, at "the earnest request of some foreign students."[63] Whether the "foreign students" appealed to the curators on their own initiative or were asked to do so by Boerhaave or his supporters was not clear.

Boerhaave successfully used his appointment as lecturer as a gateway for an appointment to a chair on the medical faculty. Within a year of Boerhaave's appointment as a medical lecturer, Stadholder-King William was killed in a riding accident, and the Leiden curators were able to fill the three vacant medical chairs at Leiden with experienced men.[64] Suddenly, with five professors on the medical faculty, there was no need for Boerhaave's services. Yet, because he had a three-year contract, he continued to teach the *institutiones medicae*, and he refused to give up his chemistry teaching, which he began in January. As had always been the case at Leiden, the success of a professor was measured primarily by (the perception of) how many students he attracted to his lectures and, by extension, to the university and town. When, in early 1703, the curators of the University of Groningen offered Boerhaave a chair in medicine, the Leiden curators and burgemeesters countered with a promise that he would be appointed to the next medical chair that should become available, again citing "the influx of native & foreign students" brought to Leiden by his courses. (The medical faculty was already composed of five professors, the maximum number allowed in any faculty by the university statutes.)[65] A chair on the medical faculty finally came open after six years, when Peter Hotton died during the exceedingly cold weather in January 1709. Boerhaave was appointed to the chair of botany and medicine at the next meeting of the curators and burgemeesters on February 18, despite the fact that he had neither taught nor been formally trained in botany. When news of Hotton's death spread, several experienced botanists inquired about the position, but the majority of the curators and burgemeesters agreed that it was more important to keep Boerhaave in Leiden than it was to have an experienced botanist.[66] Boerhaave entered into his new position with characteristic zeal and immediately began to rebuild and recatalog the botanical garden.

As suggested earlier, the main reason for Boerhaave's success was that he was an extremely productive teacher. In fact, he seemed to be much more productive than his medical faculty colleagues. Anecdotal accounts describe him as rising early, lecturing to medical students all morning, and seeing patients all afternoon.[67] As dictated by his original appointment as lecturer, every fall and spring he taught the *institutiones medicae*, the basic medical theory course at Leiden covering physiology (or the "animal economy," in Boerhaave's course), nosology, signs of diseases, the promotion of health (i.e., regimen), and the cure of diseases. He expanded this last section into another complete course of its own that he called Historia Curationenque Morborium, which he taught under his mandate to teach the *institutiones*. This new course, similar to Leiden's *praxis medica* course, comprised an ordered list of diseases and their best treatment. Boerhaave later published textbook versions of each of these courses: his *Institutiones Medicae* (first edition, 1708) and *Aphorismi de Cognoscendis et Curandis Morbis* (first edition, 1709).[68] In addition to these core medical courses, he taught a chemical lecture course during the winter term (roughly November through March). Beginning in January 1703, he augmented the lecture course with a series of chemical demonstrations in which he exhibited basic chemical techniques and recipes. He also offered private medical courses on specific subjects. In 1702 he taught a course titled De Ortu Hominis, which examined human reproduction and embryonic development, and in 1703 and 1704 he presented a course, De Sectis Medicorum, in which he reviewed and critiqued the primary claims of all of the major approaches to medicine.[69] During any given term, he offered at least two different courses.

MEDICAL PHILOSOPHY

Boerhaave attracted students and garnered the support of the university community by the way he *presented* his medicine. His medical system was a synthesis of modern experimental and "mechanical" medicine and natural philosophy, but he portrayed it as deriving from an ancient, empirical, "Hippocratic" medical tradition. This was the main argument of his first academic oration, *Oratio de Commendando Studio Hippocratico*, given after his appointment as a medical lecturer in 1701. As the title of the oration suggests,

Boerhaave made several arguments, ostensibly directed at medical students, regarding why the study of the Hippocratic Corpus was useful for the physician. Except for the corpus itself, however, all of the medical authors that Boerhaave recommended to students in the oration were modern. The most numerous group were the anatomists, beginning with Andreas Vesalius and including William Harvey and Boerhaave's own teacher, Charles Drélincourt. He also praised the Italian "iatrophysicists" Marcello Malpighi, Giovanni Borelli, and Lorenzo Bellini; the microscopic investigations of Antoni van Leeuwenhoek, Henry Powers, and Robert Hook; the chemistry and physiological experiments of Robert Boyle; and the "second Hippocrates," Thomas Sydenham.[70] For Boerhaave, the point of this exercise was to suggest that these diverse modern medical writers were part of the Hippocratic revival, which had enamored medical faculties, including the University of Leiden's, since the late sixteenth century.[71] Built upon these sources, his medicine, he suggested, was "Hippocratic" too.

Boerhaave argued that the writings of Hippocrates were essential for the student of medicine not only because they contained reliable and useful accounts of diseases, but also because Hippocrates provided an exemplary model for making medical knowledge to be emulated by students. In Boerhaave's reading of Hippocrates, the "father of medicine" was a strict empiricist who collected and organized medical observations without regard for theoretical assumptions. To understand the cause of a disease, he argued, one must "acquire a most accurate knowledge of its properties and forces, each considered separately," which may "only be perceived by our mind through their effects which are open to sense perception." Everything related to "causes, accidental details, or effects" of a disease must be examined and described in a clear manner—"so clear that no sane mind may doubt it or use rational arguments against it"—before one can arrive at conclusions regarding good or ill health. Yet to compose a reliable account of the properties of a disease, one must be free of theoretical bias: "But who will be able to follow Nature as his sole guide when investigating facts? Who will never go astray? Who will always avoid uncertainties? Only he, in my opinion, who is free from all sectarianism, unfettered by any preconceived ideas, devoid of all leanings toward prejudice; he who merely learns, accepts, and relates what he actually sees."[72] Boerhaave contended that Hippocrates was such a man and that his writings reflected the clarity and freedom from bias essential for progress in medicine.[73]

Boerhaave posited, however, that empirical observation was not enough to ensure reliable medical claims, but that Hippocrates made his practices reliable through cautious and repeated testing of his treatments. For Boerhaave, the physician's task was to remove sickness from the patient, but to succeed in this goal consistently and reliably, the physician must follow Hippocrates and assist "the natural process of the disease, rather than disturbing it by foolhardy intervention." To develop a reliable treatment for a given disease, the physician must first understand the course of the disease and then find remedies that aided the body in expelling the diseased matter from itself. But Hippocrates discovered and developed such remedies only after years of practical experience. Boerhaave argued that Hippocrates never taught a remedy to his students or recorded its preparation in writing until "he had proved [it] to be successful in a thousand cases," and even then he presented the remedy with "cautions and warnings" in order to prevent the inexperienced physician from misusing it.[74] Such safeguards not only protected the Hippocratic physician's patient from improper treatment, but they also ensured the reliability of Hippocrates's knowledge claims. As he counseled his students, "he who consults the pronouncements of the Greatest Physician need not be afraid of the dangers of ignorance, of anxieties caused by indecision, or of errors consequent upon recklessness; this doctrine is brilliant, pure, purged of all stains of error."[75] The message was that Hippocrates's methodology for medicine was authoritative.

Boerhaave's strong advocacy of Hippocrates was a means of validating certain modern medical practices. Indeed, his reading of Hippocrates as a champion of empirical medicine was an early modern one originally designed to undermine scholastic medicine. In Leiden, Boerhaave's choice of Hippocrates as the symbol for his medical program had additional rhetorical power, because according to the original graduation requirements for the medical faculty, degree candidates were only examined on the practices and opinions of Hippocrates and Galen.[76] But Boerhaave's account of Hippocrates's method of knowledge making—value-free observation, the confirmation of treatments by repetition, and a sober and cautious approach—was a clear imposition of modern medical practice onto an authoritative medical figure. Boerhaave's "Hippocratic method" could easily be attributed to Francis Bacon or Thomas Sydenham, and in fact, he probably arrived at this interpretation of Hippocrates in part by reading their works.[77] He justified this move in his oration by presenting a brief biographical sketch of Hippocrates himself. In this sketch, he presented Hippocrates as collector of observations, someone who traveled widely to observe epidemics and the diseases common to certain locations, who sent his students and relatives on similar trips, and who studied the disease accounts in the ancient Egyptian, Babylonian, and Chaldean (i.e., neo-Babylonian) medical texts.[78] Boerhaave tailored his account to emphasize the group of Hippocratic writings, such as *Epidemics*, *On Airs, Waters, and Places*, and *On the Ancient Medicine*, that he accepted as genuine and that were, in effect, those he could interpret as supporting his view of Hippocrates as an empiricist.[79]

By associating the authority of Hippocrates with his own empirical methods, Boerhaave also crafted an account of the history of medicine, which he could use to discredit other medical practices that had fallen out of favor in the Leiden medical faculty. According to Boerhaave, medicine reached its highest state with Hippocrates, but through the straying of subsequent practitioners from the method of experience, medicine declined over the centuries. The cause of medicine's fall was the improper application of metaphysical speculation. Boerhaave asserted that "medicine degenerated from painstaking observations to the assertions of philosophers, from precepts of nature to garrulity, from the utterances of Hippocrates to wanton fantasies." He first criticized the "so-called chemists," who had promised great cures but in reality were either deceitful quacks and mountebanks or people who had deceived themselves by their adherence to an unreliable method derived from their expectations and ignorance.[80] His main critique, however, was directed against the Cartesians, who comprised the most influential medical and philosophical "sect" in the Dutch Republic. He described the Cartesian approach to medicine as follows:

> [They] prefer to take general principles of nature, like matter, movement, and the shape of corpuscles, as their starting point—*a priori*, as they call it—in demonstrating the essence of health, of diseases, and remedies. Not at all embarrassed by a multitude of unproven assumptions, a lack of factual data, they set up a vague hypothesis; then, having posited this, they draw some rather sweeping conclusions via analytical thinking and using a seemingly plausible reasoning; afterwards they apply this to the facts and do not hesitate to invent rules for healing based on this kind of ingenuity. But now it has become evident that they are disappointed in their ambitious expectations when they try to use these figments of their imaginations in practical medicine, to the great detriment of their patients.[81]

The Cartesian physicians, as Boerhaave described them, employed an approach opposite to that required of the proper Hippocratic physician: they made claims based on unproved assumptions, not empirical observation.

Boerhaave filled the theoretical space vacated by the Cartesians with the methods of "modern" mathematics. He acknowledged that while Hippocrates provided an essential guide to practice and treatment, he lacked the understanding of the anatomical and physiological structure of the body necessary for a complete medical system. For this aspect of medicine, the physician must turn to the discoveries of modern practitioners, and Boerhaave had well-defined ideas on which practitioners were reliable and which were not. Reliable practitioners followed what Boerhaave called the "method of mechanics" or, alternatively, the "method of the geometers." He contended that the method of mechanics was a model of scientific practice that mathematicians and mechanics—here meaning engineers, mechanical craftsmen (such as instrument makers), and experimental practitioners of physico-mathematics—followed in pursuing their work. These "geometers" modeled (or constructed) known mathematical entities (i.e., geometrical forms or numerical magnitudes) on phenomena they observed empirically or measured with instruments, and then, by applying mathematical rules and relations to their models, they drew reliable conclusions regarding the nature of the phenomena they observed.[82] Thus, Boerhaave's "method of mechanics" was ideally both empirical and mathematical. First, one collected reliable observations according the rigorous empirical methods, and then one employed the "rules of mechanics" to arrive at conclusions based on those observations. He asserted that knowledge constructed in this manner would be both reliable and practical. In contrast with the claims of the Cartesians, Boerhaave asserted that by applying the method of mechanics, "facts would determine the argument and not the other way around. Thus, conclusions would not be vague, adaptable at will to all problems raised, and abstract, but rather well-defined, exact, and descriptive of a particular phenomenon as it really is."[83]

For Boerhaave's application of mathematics to medicine to be successful, he had to show how his new approach supported traditional institutional values in Leiden. We have seen how he portrayed his medicine as thoroughly "Hippocratic." Boerhaave deemed that the prospective physician needed to be taught mental discipline, social responsibility, and skeptical reserve in order to engage in proper medical practice. Mathematics was essential for medical education—not only for its applications but also as a tool for disciplining one's mental facilities. In mathematics, especially geometry, the student "learn[ed] through excellent precept to distinguish the clear from the obscure, true from false, and to equip his mind with prudence." The "prudence" of geometry instilled in the mind a maturity that enabled the student to grasp complex concepts clearly, "sharpened" his observational abilities, and limited his theoretical "fancies."[84]

In addition to the discipline of mathematics, the true physician must also adopt the moral character of the ideal physician, Hippocrates. For Boerhaave, the careful collection of observations, unbiased analysis of those observations, and cautious approach toward potential treatments marked Hippocratic practice. This practice, however, was grounded in a moral stance that placed the interests of the patient above those of the physician. Following the Hippocratic model, a physician should always be truthful, open, modest, and neither "secretive about unfavorable outcomes; nor boastful of his successes."[85] Boerhaave's Hippocrates did not practice medicine for his own glory but rather out of a desire to serve mankind, and for Boerhaave, the suppression of self-interest for the good of the community reflected the deeper moral character that guided Hippocratic medical practice. Toward the end of his *Oratio de Commendando,* Boerhaave attested that physicians who had mastered the method of Hippocrates were likely to become rich because of their skills, but he implored his students to follow a higher morality: "You . . . who have higher aspirations, who consider virtue to be its own and only reward, who seek after a science which, when put to practical use, is salutary to the sick, useful to your country, fills your parents with pride, and is a credit to yourselves—you should train yourselves in the work of this man [Hippocrates]."[86] Indeed, following a commonly repeated theme among Calvinist *predikanten,* Boerhaave held that the desire for money—avarice and cupidity—was a form of spiritual pollution that clouded one's reason.[87] Only a physician whose mind was disciplined by mathematics and guided by a desire to serve others was a trustworthy medical practitioner.

Boerhaave's model physician possessed the same discipline and ethics as the model Calvinist, and Boerhaave underscored the connection between medicine and religion overtly in his orations. He believed that medicine was a religious "calling" through which one served the Christian community by acting as God's instrument. The physician's aim of curing the human body to ease suffering was analogous to the preacher's aim of bringing peace to the human soul. Boerhaave therefore employed religious language to describe the merits of his medicine. For example, in extolling the cautious and meticulous empirical method that he attributed to Hippocrates, Boerhaave remarked: "There is no other way, nor will any other be found which does not lead to the perdition rather than to the salvation of mankind. Only this one is admirable, useful, nay, almost divine."[88] He underscored the "divinity" in medicine by situating medical research within a natural theological context. In his 1703 oration, *De Usu Ratiocinii Mechanici in Medicina,* Boerhaave portrayed God as both a "master mathematician" and "most expert mechanic," who conceived and created the structure and mechanisms of the human body. By revealing the body's structure, the anatomist and physician simultaneously revealed God's divine artifice as testimony of His presence and majesty.[89] Boerhaave's medicine and plan for medical education thus reflected the same Calvinist religiosity that marked his public behavior and character.

WHY BOERHAAVE?

The reasons for the acceptance and success of Boerhaave's program for medical reform are varied. In his first academic oration, Boerhaave linked his program for the reform of medicine to established pedagogical traditions and religious values in Leiden. First, he presented his medical method as derived from the Hippocratic tradition, which had been the traditional basis of the medical faculty's curriculum. Second, he presented his new curriculum as an enterprise that shaped young physicians into good citizens though the discipline of mathematics and by emulating the model of the ideal, Hippocratic physician. In both cases, Boerhaave made modern experimental and mathematical approaches to medicine, upon which his curriculum was based, orthodox in the sense that he portrayed his reforms as supporting the Leiden pedagogical tradition and the doctrines of the Reformed Church. Boerhaave's linking of his method with medical tradition and Protestant values connected the interests of various factions, broadly speaking, within the university and Dutch society. The message that he broadcast to the curators and burgemeesters was that he planned to train physicians to have good moral character, a disciplined work ethic, and the desire to become servants of the community and state. For the professors and students in the liberal arts and medical faculties, Boerhaave had found a way to pursue the new approaches in natural philosophy and medicine with the full blessing of the community.

In effect, Boerhaave's success as a medical lecturer at Leiden depended on his ability to gain the support of the curators and, to a lesser extent, the students at the university. He accomplished this task by presenting himself and his medical method as derived from the local social and intellectual culture. He continued this rhetorical strategy throughout his tenure at Leiden. Take, for example, his remarks at the close of his third university oration, *Oratio qua Repurgatae Medicinae Facilis Asseritur Simplicitas*, which he presented in 1709 on the occasion of his acceptance of the chair of botany. In the oration Boerhaave thanked all of those in the audience who had helped him obtain his chair. Of course, he thanked the curators and the students, but he also named four professors who were current members of the faculty. He praised Jacob Gronovius for helping him "to polish my unfit style of writing," Wolfred Senguerd for his "most clear and kind teaching on the subject of philosophy," and Johannes à Marck, for his lecturing on theology. He reserved the greatest accolades for Burchard de Volder, whom he called "the wisest of men, unequaled in the spotless purity of your character and virtue!"[90] Boerhaave included these panegyric praises for his teachers to remind his audience that he was a product of Leiden University. Whatever they may have thought about his claims and views on medicine, which he had just expounded for the third time, he was not radical or threatening. He was one of them.

CHAPTER TWO

Didactic Chemistry in Leiden

Soon after he received his medical degree in 1693, Herman Boerhaave established a private medical practice in Leiden. In his spare time, he conducted chemical experiments in a small laboratory that he had set up in a shed near his house. Many of his experiments addressed a burning chemical problem for him: the transmutation of metals. By 1696, Boerhaave was working on the preparation of "sophic" mercury, which was, according to the mercurialist school of alchemy, a necessary ingredient to be used in preparation of the philosophers' stone. Mercurialist texts described this mercury as an extremely pure, subtle, and structurally uniform variety of ordinary quicksilver, whose particles had the power to penetrate the bodies of other metals, dissolving them into their constitutive principles or prime matter. When used to dissolve gold, this sophic mercury released the gold's generative power, which the chemist could then manipulate to transmute other metals.[1] To create this sophic mercury, Boerhaave was working through the texts of Eirenaeus Philalethes, perhaps the most prominent alchemical authority in the second half of the seventeenth century.[2] The Philalethan corpus contained a recipe for sophic mercury, but this recipe was not presented openly in Philalethes's texts. Rather it was concealed through *decknamen* (i.e., metaphorical names for ingredients and processes) and allegorical representations and was dispersed through several different texts.[3] Where could Boerhaave turn for assistance in deciphering this puzzle?

Luckily, Boerhaave owned a copy of the lecture notes from a course given by a previous professor of chemistry at Leiden, Carel de Maets (1640–1690). Among the processes, which De Maets described, was one for "mercury of the philosophers according to the mind of the anonymous Philalethes." Just as with the other recipes he outlined in his course, De Maets did not explain the theory behind the process (at least it was not recorded in the lecture notes), but he revealed the process in terms of mundane chemical language and principles. The recipe called for the creation of an amalgam of "regul[us] [antimon]ii [mart]ialis" and "[lun]ae cupell[ae]." This amalgam was then fused with "vulgar mercury." This second amalgam was ground with a mortar and pestle and placed in boiling water until its "blackness" was removed, and it was "strongly whitened inside and outside." Finally, the mercury was distilled out of the white amalgam through "successive ascensions" in an alembic.[4] The operations and ingredients in De Maets's recipe were fairly common to metallic assaying: "cupellation" was a common method of purifying silver described in chemical and assaying manuals, as was the "washing" of amalgams with water to remove scoriae. "Regulus antimonii martialis" was produced by reducing antimony ore (antimony sulfide or stibnite, simply called "antimony" in seventeenth-century terminology) to metallic antimony through an operation that used iron (the "martialis") as the reducing agent.[5]

What Boerhaave found in his two sources were two different modes of chemical discourse. Eirenaeus Philalethes was a fictitious character invented by the American alchemist George Starkey (1628–1665).[6] Whereas Starkey, via Philalethes, treated sophic mercury as a grand secret to be hidden from the common masses, De Maets presented Philalethes's mercury as an ordinary chemical preparation, albeit one with curious powers and uses. These two modes of presenting chemical knowledge reflected the different contexts in which the authors of these texts deployed them. Starkey parleyed his supposed acquaintance with the adept, Philalethes, to gain patronage from Robert Boyle and members of the Hartlib circle during the 1650s. The enigmatic, obtuse nature of Philalethes's texts allowed Starkey to position himself as the best source of information and skill for those interested in Philalethes's secrets.[7] De Maets, however, as professor of chemistry, established his professional status through the open communication of knowledge. The fact that Philalethes's sophic mercury could be considered an alchemical secret by some only served to reinforce De Maets's academic commitment to the open dissemination of knowledge and, perhaps, titillate his young students.

During his formative years in chemistry, Boerhaave interacted with both of these modes of chemical practice and communication. He was interested in the secrets of alchemy as a means of revealing the properties of chemical species and powers of nature. But as an academic, he was also committed to modes of open communication. This commitment derived from his training in medicine and natural philosophy at Leiden. This training, however, also induced Boerhaave to question the didactic tradition in chemistry. From the perspective of most university physicians, chemistry was a form of pharmacy and therefore was traditionally under the purview of apothecaries and other chemical artisans. As such, it was not a part of the core of medical education. Didactic chemistry courses and textbooks did not fit into established modes of pedagogical practice in the medical faculty, which defined theoretical precepts and then examined how those precepts were to be applied. Chemistry courses, by contrast, were seen as collections of practical recipes with limited theoretical content. Despite the interest in chemistry by medical students and even a few medical professors, Leiden's professors of chemistry were unable to achieve full academic status for themselves or their field because of their perceived lack of philosophical rigor.

This argument challenges Owen Hannaway's interpretation of the didactic textbook tradition. Hannaway argued that Andreas Libavius's *Alchemia* (1597) founded the tradition of chemistry textbook writing and was dedicated to the open communication of chemical recipes and ideas. This act was an important event in the history of chemistry, because, according to Hannaway, the didactic tradition defined itself in opposition to the esoteric and secretive modes of the Paracelsians and alchemists.[8] Subsequent historical work, however, has significantly altered this understanding of the didactic tradition by showing how practicing chemists adopted their courses to fit local contexts and philosophical styles.[9] Later textbooks were used as vehicles for the spread of Paracelsian remedies and practices as well as alchemical knowledge.[10] The significant point here, however, was that Hannaway, and more recently, Bruce Moran, argued that Libavius deployed academic modes of argument and organization in his chemical work. The *Alchemia* was part of his attempt to make chemistry into an academic art.[11] Libavius's project was, in part, a failure. His textbook was co-opted by artisan chemists, who turned his academic presentation of chemistry toward the needs of the practical arts. Chemical textbooks and courses after Libavius were designed to help apothecaries pass their licensing exams, record chemical recipes and operations, and entertain an increasingly interested public. Ultimately, Boerhaave completed Libavius's reform of chemistry by remaking it into an academic art.

This chapter examines the rise and fall of didactic chemistry at the University of Leiden from the founding of a chemical laboratory and professorship to the decline in chemistry teaching by the 1690s. The story begins with a discussion of the origins of didactic chemistry with Libavius, its development in the hands of apothecaries, and then its ambiguous position at the University of Leiden. The chapter concludes with an examination of Herman Boerhaave's education in chemistry and his interaction with the didactic tradition. Boerhaave valued didactic chemistry as a body of chemical operations, skills, and concepts that any competent chemist needed to master. It did not, however, provide him with a methodology to investigate the properties of chemical species, which was his primary interest in chemistry. As a result, he turned to other chemical traditions—transmutational alchemy, chemical medicine, and experimental natural philosophy—to inform his chemical investigations.

THE RISE OF DIDACTIC CHEMISTRY

As late as the last decade of the sixteenth century, there were no books one could read or courses of study one could undertake that synthesized all of the topics and practices that could be called "chemistry." There were books on craft traditions in chemistry, such as the anonymous *Probierbüchlein* on assaying, Hieronymus Brunchsweig's *Art of Distillation*, and various pharmacopoeias compiled by apothecaries. Another category included books, which we might describe as "alchemical," on the transmutation of metals and other arcana, and composed by chemical philosophers and adepts. Finally, there were the books on chemical medicine composed by Paracelsus and his followers, who deployed chemical operations to make "spagyric" medicines, supposedly superior to traditional remedies. In each of these cases, however, "chemistry" was something that one applied to various crafts or arts in pursuit of the goal of that art: testing the quality of a metal, transmuting it into a precious metal, or fashioning it into a remedy to cure the sick. But what was chemistry in itself? Indeed, this was the question that Andreas Libavius (ca. 1555–1616), Rothenberg town physician and inspector of schools, asked in the 1590s. As Libavius understood the situation, all of these texts had something to say about chemistry, but few addressed "chemistry" as a proper art in itself with its own content, order, and principles.[12]

Arguably, Libavius's *Alchemia* of 1597 was the first textbook solely about chemistry. Libavius composed the *Alchemia* and a battery of subsequent chemical writings to remove chemistry from its applications and give the field a classical, humanist foundation as an "art": "a system of precepts exercised together towards some useful end."[13] As such, he presented *al-*

chemia as an ordered collection of precepts taken "from various authors" and arranged according to a pedagogical method, which "ought to govern the disposition of the sciences."[14] Libavius's work was inspired by the dialectic of Petrus Ramus (1515–1572), who devised a universal method to order any science or art in the most pedagogically expedient manner. Ramus's "method" proceeded by positing a general definition for the field under scrutiny and then defining two topics, which were incorporated within the first definition. These topics where then divided into two more topics, which were also defined. This process of definition and division proceeded from general to specific topics until all of the precepts for the art in question had been defined. The final product was, according to Ramus and his followers, the "natural" pedagogical order of the field, being clearly presented, easily apprehended, and logically persuasive.[15] Following this model, Libavius began his textbook with a definition of *alchemia*: "Alchemy is the art of perfecting magisteries, and of extracting pure essences from mixts by separating them from their bodies."[16] He then divided alchemia into two topics: *encheria*—general techniques and classes of operations—and *chymia*—the chemical species prepared through the operations. Each of these topics was in turn defined and divided, continuing until all the precepts included in *alchemia* had been specified.[17] The act of defining and ordering the concepts and operations of *alchemia* into pedagogical topics was an immensely important event in the history of chemistry, because it created an organized field of knowledge where there had been none before. In Hannaway's words, "Libavius called chemistry into being simply by demonstrating that it was teachable."[18]

Libavius's pedagogical organization and style became the inspiration and model for later chemical textbooks during the seventeenth century.[19] These textbooks comprised what Hannaway called the "didactic" tradition in chemistry, and the first ones were generally structured according to the form established by Libavius and his first imitator, Jean Beguin (fl. ca. 1595–1615). Beguin was a Huguenot apothecary who offered chemical instruction at his Parisian shop to help young apothecaries pass their guild examination and also to promote new spagyric or Paracelsian remedies, which he favored.[20] In 1610 Beguin's students published a slim, anonymous volume of notes compiled from his course and titled *Tyrocinium Chymicum*, which he himself reworked and published under his own name in 1612. Beguin used Libavius's *Alchemia* as a model to structure his chemistry course, as shown by the many places in the *Tyrocinium* where the order and definitions of topics replicate those found in Libavius's work.[21] The popularity and importance of the *Tyrocinium* cannot be underestimated. Pirated editions of the first, anonymous version appeared a year after its 1610 publication, and further pirated editions of Beguin's 1612 version appeared in the same year. In response to the demand for the *Tyrocinium*, Beguin issued a significantly enlarged French version, under the title *Elemens de la chimie*, in 1615. Versions of the Latin *Tyrocinium* and the French *Elemens* continued to be published throughout the seventeenth century. In effect, Beguin's work functioned as the structural foundation for the didactic tradition, as editors of subsequent editions of the *Tyrocinium* or *Elemens* added their own material to the extant text. For example, by 1625 the "sixth edition" of the *Tyrocinium* published by "Christopher Glückrad," a pseudonym for the Marburg public professor of *chymiatria*, Johannes Hartmann (1568–1631), had grown to 392 pages, a threefold increase from Beguin's 1612 edition.[22]

The seventeenth century saw a proliferation of chemical courses and textbooks, along with new institutions to support didactic chemistry. The number of didactic textbooks published was staggering. T. S. Patterson listed forty-two separate editions of Beguin's *Tyrocinium* or *Elemens* alone in the Ferguson Collection at the University of Glasgow.[23] Lynn Thorndike identified twenty-four separate textbook titles in his monumental *History of Magic and Experimental Science*, and this number does not take into account multiple editions.[24] As with Beguin's textbook, the most popular works appeared in numerous editions. At the end of the seventeenth century, for example, Nicolas Lemery (d. 1715) expanded and reissued his *Cours de chymie* ten times himself from 1675 to 1713, and at least another twenty unauthorized editions, including translations from French into Latin, German, Dutch, Italian, Spanish, and English, appeared by 1757.[25] The hundreds of chemistry textbook editions represent only a fraction of the number of chemistry courses offered during the century. Most of these courses were offered at the shops of artisan chemists, but some were also taught at salons, the houses of gentlemen, or public establishments as chemistry became a subject of interest among the educated elite. New institutions for didactic chemistry were also founded, the most famous of which was the government-funded course taught at the Jardin des plantes in Paris. The regular, public courses, which began in the late 1640s, made the Jardin the center for chemical instruction in France, and three of the first chemist lecturers, William Davidson (fl. 1648–51), Nicaise le Fèvre (fl. 1651–60), and Christolphe Glaser (fl. 1660–72), published textbook versions of their courses.[26]

The popularity of didactic chemistry signaled a change in the traditional pedagogy of chemistry. In the seventeenth century (and before), novice chemists were trained through apprenticeship: submitting to the discipline of a master chemist and learning techniques by observing and doing. This training may have been augmented with the study of manuscripts or printed handbooks that contained recipes, described techniques and apparatus, and defined terms, but the aim of this training was practical know-how.[27] The didactic presentation of chemistry augmented the established mode of training with organized lectures and book study. Chemists embraced this new mode of training because it was an efficient method for organizing and transmitting basic knowledge. Within a few months, novice chemists attending a chemistry course could learn basic terminology, concepts, and techniques; often the lectures were accompanied by practical demonstrations, which served to connect the sensory-dependent aspects of chemical work with their textual descriptions.[28] After completing the course, students would have a useful reference for later examinations in their course notes or a purchased textbook. As Bruce Moran has shown, making the connection between practical skills and recipes and their textual representation (in the form of students' notes) was an explicit aim of Johannes Hartmann's didactic course at Marburg.[29] Although the apprentice still needed laboratory training to become a competent, practicing chemist, didactic courses became an important tool in the chemist's overall preparation.

The greatest impact of didactic chemistry, however, was its ability to transmit chemical knowledge beyond the circle of artisan chemists. The same pedagogical principles that made didactic chemistry appealing to novice chemists made it accessible to nonchemists. By the middle of the seventeenth century, the audience for didactic chemistry had expanded to include physicians, natural philosophers, writers, artists, mechanics, and gentlemen savants. In Paris, some chemistry courses were seen as social events, as attested by Bernard Fontenelle, who claimed that even "ladies swept away by fashion" attended Lemery's courses.[30] These nonchemists brought chemistry, which was once a specialized craft, into the public discourse on natural philosophy, medicine, and technology.

For many chemists and nonchemists alike, didactic textbooks defined what chemistry was: its aims and goals, its acceptable modes of practice, its potentials and limits, and its place in the hierarchy of disciplines. So what was chemistry in the didactic tradition? Beguin's *Tyrocinium* and the didactic tradition in general presented chemistry as a craft practice or "art." For Beguin, chemistry performed an ancillary role to medicine by processing crude medicinal substances and making them into palatable remedies. This aim was stated clearly in Beguin's definition of chemistry at the start of his work: "Chemistry is the art of separating [*solvendi*] natural mixed bodies and recombining [*coagulandi*] the separated parts in order to produce [*concinnanda*] more agreeable, more healthful, and safer medicines."[31] But the chemist was not a physician. Beguin accepted, at least in print, the traditional place of the apothecary in the medical hierarchy and indicated that the chemist's role in medical practice was to fill the prescriptions of physicians. A well-trained chemist produced spagyric medicaments that were superior to the nauseating and ineffective remedies of the "vulgar." To achieve these superior medicines, the spagyric chemist possessed both theoretical knowledge and practical skill, the combination of which made chemistry a true "art." Despite Beguin's claim that he possessed theoretical knowledge, the chemist was not a philosopher either. He lamented those practitioners who neglected the practical part of the art and resented the fact that many persons believed that "alchemists" only meditated on the arcane mysteries of the philosophers' stone.[32] For Beguin, the chemist was an artist whose practice was guided by a rational understanding of his operations, but whose professional interests stopped at the fabrication of chemical products.

Indeed, Beguin's overwhelming concern with the practical understanding of operations rather than the philosophical causes of chemical change was a dominant feature of didactic chemistry. Libavius strove to portray chemistry as a practical art, and as such his primary aim was to describe the operations and products of the art rather than make arguments about the nature of matter or how chemical operations could reveal properties of matter. The structure of Beguin's *Tyrocinium* also reflected this aim. He divided the text into two main sections that examined chemical operations and chemical products, respectively. The first section, comprising book 1 of the *Tyrocinium*, categorized all chemical operations under one of two headings: separation (*de solutione*) or recombination (*de coagulatione*), the two primary actions of the chemical art in Beguin's definition of chemistry. Each chapter defined a specific type of operation (calcination, filtration, distillation, etc.) and described how the operation affected the bodies upon which it worked. For these definitions, he often used theoretical concepts and language borrowed from earlier chemical sources. He defined "calcination" as the way to pulverize a substance through fire by depriving it of the "humidity" consolidating its parts. This definition was a direct quotation from the *Summa Perfectionis* of "Geber," a medieval work on metallurgy and transmutational alchemy.[33] The quotation, however, did not serve as the basis for further philosophical discussion but rather as an organizational locus, which allowed Beguin to describe specific operations that he classified as types of "calcination." The aim was to depict the tools and techniques at the chemist's disposal, not examine the natural principles behind their operation. This practical aim was emphasized at the conclusion of book 1, where Beguin discussed the construction and operation of chemical furnaces, vessels, and lutes.[34]

The second and largest section of the *Tyrocinium*, comprising the last two books, examined the products of chemical operations. The first chapter of book 2 offered a list of rules (*canones*) for conducting distillations, but each subsequent chapter focused on a class of chemical product (waters, spirits, tinctures, oils, extracts, etc.). The individual chapters began with a definition of the chemical entity named in the chapter title but quickly proceeded to recipes detailing how to synthesize specific products and included a discussion of their uses. The classification system outlined by the chapter headings reinforced the practical nature of this enterprise. Products were not grouped together according to their substance but rather through operational categories related to their synthesis and observed characteristics within the laboratory. The chapter "On Spirits," for example, described recipes to make rectified spirit of wine, spirit of tartar, spirits of niter and vitriol, and the *spiritus sylvestris* of Paracelsus.[35] The only characteristic that unified these products as a class of chemical entity was that they were "volatile" substances (i.e., collected as a vapor) obtained through similar processes of distillation. In a few chapters Beguin did seemingly group products by substance, such as in the two chapters on antimony and mercury. These chapters highlighted the fabrication of Paracelsian, metallic remedies, and Beguin separated them from earlier chapters to emphasize the special operations needed to prepare metals and, also, to signal their special medicinal uses. Even in these chapters, the bulk of the text concerned recipes for medicines, not a discussion of the natural properties of these metals.

Almost all chemistry textbooks produced in the seventeenth century followed the structure and practical emphasis of Beguin's *Tyrocinium*. Most subsequent didactic textbooks were composed of two books or sections: the first, shorter book contained a definition and, perhaps, etymology of the term "chemistry," and a discussion of the types of operations, equipment, and whatever system of chemical principles the author advocated. The second and always larger book described chemical products and focused on recipes. Even as the specific content of each section evolved over time and varied between specific authors and locales, the basic organizational structure and emphasis on operations and products remained. For example, at the beginning of the eighteenth century, the tenth edition of Nicolas Lemery's *Cours de chymie* (1713) followed this standard two-section format. This first section was a 68-page "prelude," which presented the etymology and definition of *chymie*, an extended discussion of the chemical principles (salt, sulfur, mercury, phlegm, and earth), a description of chemical furnaces and vessels, and a short glossary of terms. The second and main part of the text contained 880 pages of recipes and products. Lemery had dropped Beguin's operational taxonomy, preferring to organize his operations based on the substances or "simples" upon which the operations worked. The text divided the simples into the three kingdoms: mineral, vegetable, and animal. The section on each kingdom contained individual chapters devoted to specific simples (e.g., gold, silver, tin, etc., under minerals). In the chapters, Lemery defined and described each simple and enumerated the recipes that focused on that simple. Although he did not organize his book around types of operations (this approach was going out of style by the eighteenth century), Lemery still placed the overwhelming emphasis on recipes, techniques, and practical applications. He described in detail the operational procedures for each recipe and included a discussion of each product's uses in a *remarques* section following each recipe. To assist apothecaries and physicians in using his text in their practice, he appended to his book an index of *vertus* exhibited by his remedies. One simply found the condition that needed to be treated or the effect generated, and Lemery identified the applicable remedies.[36]

Despite the commercial success of didactic chemistry—chemistry textbooks sold well and courses were often well attended—chemists did not make significant headway into the university. Because the chemist was seen as an artisan whose job was to fabricate the prescriptions of physicians, university faculties did not consider chemistry an appropriate topic to include in the university curriculum. There were exceptions. Johannes Hartmann taught courses in *chymiatria* at the University of Marburg, and a tradition of chemical instruction developed at the University of Jena.[37] Additionally, in university towns medical students and faculty often attended didactic courses offered by apothecaries in their shops or other public venues. Peter Staehl's courses, presented to students and luminaries like Robert Boyle, around the University of Oxford during the early 1660s, were examples of this practice.[38] These courses, however, were neither required nor encouraged by any university. When the University of Leiden moved to establish a professor of chemistry, the new professor faced the same problems of status and placement within the curriculum as chemists at other universities.

DIDACTIC CHEMISTRY AT LEIDEN

The University of Leiden was one of the first universities to have regular chemical instruction. In 1669 the curators and burgemeesters of Leiden appointed Carel de Maets as professor of chemistry and established a chemical laboratory for the professor to offer courses in chemistry at the university. This event has traditionally been interpreted as the birth of the first sustained chemistry program at a university.[39] The professorship and laboratory, however, did not ensure the success of chemistry at Leiden. Thirty years after De Maets's appointment, chemistry was still a marginal subject at Leiden. The didactic chemistry, which De Maets and other chemical lecturers promoted at Leiden, did not fit easily into either the pedagogical or institutional structures of the university. As at other universities, chemistry at Leiden was seen by professors of philosophy and medicine as lacking philosophical rigor. It was a form of *techne*, which properly functioned as a "handmaiden" to more established disciplines, like medicine. As such the professor of chemistry and his field retained a low status within the university curriculum.

For the first sixty years after its founding in 1575, the University of Leiden did not offer chemical instruction as part of its curriculum. Students in Leiden who took an interest in chemistry relied on local artisan apothecaries for instruction. These chemists were often amenable to taking on university students as short-term apprentices or observers in their shops. The instruction that the student received was primarily practical: how to handle equipment, identify and work with simples (impure starting materials), and follow common recipes. From the perspective of the university, this instruction in chemistry was extracurricular. The chemist and student negotiated the terms of their relationship without the interference of the university, and the training the student received had no bearing on his university studies. Such arrangements between individual students and artisan chemists were fairly common in Leiden and persisted throughout the seventeenth century.[40] Neither the university curators nor the medical faculty professors objected to this form of instruction, because they did not see it as infringing on medical faculty courses.

In 1642 the medical faculty attempted to institutionalize this model of chemical training by enlisting a local apothecary, Niclaes Chimaer, to demonstrate the fabrication of medicaments as part of the faculty's clinical course (*collegium medico-practicum*). As part of this course, which gave medical students practical, bedside experience as part of their university training, Leiden professors took students to see patients in two wards reserved for this purpose at the St. Catherine and Cecilia Hospital in Leiden. In a reinforcement of the traditional medical hierarchy, Chimaer was charged with showing medical students how the professor's prescriptions were prepared. He was also commissioned to give general instruction in pharmacy to university students upon request.[41] No medical student, however, was required to receive instruction from Chimaer. This arrangement was designed to make things easier for a student who wished to learn pharmaceutical preparations on his own initiative.

Franciscus Sylvius (or François de le Boë, 1614–1672) was the first medical professor to integrate chemistry into his teaching at Leiden. Sylvius was a proponent of chemical medicine, which stemmed from a Paracelsian tradition that sought to understand human health and disease as products of "chemical" action. His chemistry, however, was not the chemistry of the artisan apothecaries, because he was primarily concerned with constructing a chemical philosophy with which he could explain human physiology and disease. Sylvius modeled physiological processes on the actions of bodies he observed in chemical operations. At the base of his notion of chemical action lay a sophisticated theory of chemical affinity. Certain chemical species acted on one another as "acids" and "alkalis," each seeking union with the other through "effervescence" or "fermentation." An imbalance or "acrimony" (*acrimonia*) in human bodily fluids led to disease, which could then be treated with an appropriate acid or alkaline remedy, returning the fluids to a state of chemical balance.[42] Sylvius became the dominant professor on Leiden's medical faculty during the 1660s, because he was extremely popular among the students. He was a zealous and enthusiastic teacher who not only lectured on his unorthodox medical theories, but also conducted daily bedside instruction and performed many more dissections and autopsies than the other professors.[43]

As part of his program for chemical medicine, Sylvius gave private instruction in chemistry to interested students. Few records remain of Sylvius's chemistry teaching, but extant evidence suggests that this instruction focused primarily on pharmacy and was associated with the *collegium medico-practicum*. He often prescribed remedies that were not in the predominantly herbal and Galenic pharmacopoeia of Leiden. In 1666, for example, the trustees of the St. Catherine and Cecilia Hospital complained to the university curators that Sylvius's unorthodox prescriptions were placing a financial strain on the town dispensary.[44] Overall, however, his pharmacy seems to have been much less radical than his medical theory. Beginning around 1670, Lucas Schacht (1634–1689), a former pupil of Sylvius's, took over his chemistry teaching and probably continued after Sylvius's death in 1672. Schacht's course, which undoubtedly was modeled on his mentor's teaching, presented a list of preparations identified and organized by their starting simples. Each preparation included a recipe and a brief discussion of its medicinal properties and uses. The vast majority of Schacht's preparations derived from common medicinal simples, such as camphor and other herbal extracts. In fact, three-fourths of his pharmacopoeia was herbal, structured taxonomically according to the part of the plant (root, wood, seed, fruit, flower) from which the simple derived. He also described several mineral and metallic remedies associated with Paracelsian, chemical medicine, including controversial preparations with antimony, but these were given no special emphasis or significance in the course. At the end of the course, Schacht described several basic chemical operations employed in pharmacy, such as the extraction of herbal essences both with solvents and through distillation.[45] Ultimately, the content of the instruction that a Leiden student received from Sylvius's or Schacht's courses was not that different from the instruction they might have received earlier from Chimaer or in any apothecary's shop. As in the past, instruction in chemical pharmacy was essentially extracurricular; pharmacy was not required, and neither Sylvius's nor Schacht's course was announced in the official *series lectionum* of the university.

Sylvius's main legacy for chemistry at Leiden was his advocacy for a university-funded chemical laboratory. He had constructed at his residence in Leiden a small chemical laboratory in which he engaged in chemical experimentation and possibly taught his chemistry courses. During his tenure, Sylvius petitioned the Leiden curators to found a *laboratorium chimicum* at the university, which would compliment the medical faculty's botanical garden and anatomical theater. Nothing came of these requests until February 1666, when the extremely popular Sylvius threatened to resign his chair over the hiring of another professor. The curators had appointed to the medical faculty Anton Deusing (1612–1666), the personal physician to the Prince of Nassau and a professor of medicine at Groningen. Sylvius intensely disliked Deusing, calling him an "orthodox quarreler," a charge not without merit. Deusing had argued, for example, that certain diseases had supernatural causes.[46] In an attempt to placate Sylvius, their most renowned professor, the curators offered to construct a laboratory and name Sylvius professor of chemistry. When Deusing died before he could accept his appointment, however, the curators seemingly dropped their offer, although Sylvius continued to advocate for a chemical laboratory.[47]

The curators found an opportunity to placate Sylvius in December 1668, when they were approached by the young chemist Carel de Maets. De Maets offered his services as a chemical lecturer to the Leiden curators after he heard the rumor that they intended to construct a laboratory at the university. De Maets had excellent credentials for this post: he had earned a degree in philosophy from Utrecht (1664), had worked in the Amsterdam laboratory of the illustrious chemist Johann Rudolph Glauber (whom Sylvius knew personally), and had taught chemistry briefly back at the University of Utrecht as an unsalaried lecturer (1668). He came to Leiden to offer his services because the curators at Utrecht had refused to construct a laboratory for him. Motivated at least in part by a desire to capitalize on the folly of their rival university, the Leiden curators opted to accept De Maets's offer.[48] In February 1669 they grandly announced that "nothing was lacking to complete the luster of the medical faculty in service of the students, but the making of medicines in the chemical manner and the taking from experiments the effects of nature." Thus, the curators resolved to construct a chemical laboratory, and they appointed De Maets "provisionally" to give "lessons in chemistry," to "hold disputations," and to "demonstrate chemical operations."[49] This first laboratory, which opened in August 1669, was constructed in a small house next to the university's botanical garden. De Maets was provided with an allowance of 250 guilders to purchase fuel (coal, peat) and small equipment (crucibles and other vessels). He was given no salary, but he was allowed to collect lecturing fees and to keep or sell any preparation that he produced.[50]

Despite this new institutional niche in the university, the place of chemistry and the professor of chemistry within the university's academic structure was somewhat ambiguous. De Maets's appointment replicated the old model of chemical instruction at Leiden, but now the artisan chemist was simply brought in-house. This attitude toward De Maets was reflected by the curator's assumption that De Maets would sell his chemical preparations to augment his lecturing fees. Initially, the curators were very pleased with De Maets, and in 1670 they promoted him to professor *extrordinarius* of medicine, a position that allowed him to announce his courses at the university and included a small salary but did not allow him to participate in medical faculty business. In 1672, after the death of Sylvius, who was perhaps his faculty advocate, De Maets was removed from the medical faculty and appointed professor *ordinarius* of chemistry in the lower-status philosophy (i.e., liberal arts) faculty.[51] De Maets was not able to obtain a full appointment on the medical faculty until January 1679.[52]

De Maets's difficulties in obtaining a medical faculty appointment reflected the general view of chemistry as a form of *techne* and not a full-fledged academic discipline. Many university academics and educated physicians saw chemists such as De Maets as "empirics," a pejorative label implying a practitioner's lack of philosophical sophistication.[53] This view shaped the institutional role of chemistry at the university. Chemistry supplied products and practical operations for use in other disciplines but did not have its own academic structure and theoretical precepts, an indication of its lack of a true academic status. In the medical faculty, De Maets's chemistry was treated as pharmacy, which fell under the heading of "treatment" within the *praxis medica* course. As shown in the pharmacy course of Schacht, one rarely discussed a remedy in terms of its "chemistry" (i.e., the natural or methodological principles behind making it) but only in terms of its recipe and efficacy in treating specific diseases. In the philosophy faculty, where De Maets was placed during most of the 1670s, chemistry was defined as a form of *physica experimentalis* or *physica practica*. Despite the historians' fanfare associated with rise of "experimental physics" in the seventeenth century, university physics demonstrations, such as those conducted by Burchard de Volder (1643–1709) at Leiden's famous *theatrum physicum*, were intended to reinforce the concepts to which students were introduced in their regular lectures on *physica theoretica*.[54] Because De Volder gave the lectures in theoretical physics, De Maets's chemistry courses, in effect, assumed the subordinate role of providing examples for De Volder's theoretical assertions. Even De Maets himself found it difficult to portray chemistry as an independent, academic subject. In the introduction to his course, he stated that as far as chemistry considered the motions of bodies, it was classified as *physica practica*. But by manipulating the natural motions of bodies artificially, chemistry "added numerous medicines, partly by undertaking pharmacy itself and [partly] by exploring the Pharmacopoeia Spagyrica, hermetica, [and] distillationia."[55] In either case, chemistry was a "handmaiden" to established university disciplines.

De Maets's adherence to the traditional method of pedagogy presented in most chemistry courses and textbooks of the day only reinforced the perception of chemistry as *techne*. Following the didactic structure of contemporary textbooks, De Maets introduced his course with a definition of chemistry, a brief discussion of the three Paracelsian chemical principles (salt, sulfur, and mercury), and a description of four types of operations (distillation, sublimation, putrefaction, and fermentation). He covered this theoretical introduction on the first day of the course and then proceeded to the heart of the matter: the presentation (and in some cases, demonstration) of approximately two hundred chemical recipes.[56] De Maets, however, made a concerted effort to make his chemistry more than just pharmacy by discussing a wide range of operations from disparate sources. He engaged with diverse practical and literary traditions from pharmacy and assaying to alchemy and natural philosophy. In addition to the herbal remedies of traditional pharmacy, De Maets also included many mineral remedies, such as a group of antimonial and mercurial preparations, which he recommended as a treatment for gonorrhea.[57] Other preparations, which had no relation to medicine, were included either as curiosities or as lucrative products in the chemist's wider repertoire. In one version of his course, De Maets showed his students two preparations for cosmetics, one white and one red. He described several alchemical operations, including one for Eirenaeus Philalethes's "mercury of the philosophers" described at the start of this chapter. He performed an "experiment" to effect color changes in solutions taken from the work of Robert Boyle and Otto Tachenius on chemical indicators.[58] In another course, he described various methods by which one polished stones for jewelry making.[59] De Maets often appended theoretical explanations to his descriptions of operations. For example, when discussing operations with metals (especially those touching on metallic transmutation), he explained the operation by employing a version of the mercury/sulfur dyad theory of metallic composition, which he appropriated from the writings of Joan Baptista van Helmont and Philalethes.[60] His sporadic attempts to incorporate chemical theory into his courses, however, did not succeed in bolstering his academic credentials.

De Maets's position at the university was further compromised by the fact that he faced competition from private chemistry lecturers in Leiden. Christiaan Marggraf (1626–1687) studied medicine at Leiden with Sylvius during the 1650s, eventually traveling to Franeker to take his medical degree. In 1659 he returned to Leiden and established a private medical practice. At this time, he petitioned the academic senate of Leiden (composed of professors) for permission to give private lectures in medicine and chemistry with the university's sanction. His request was rejected, but Marggraf proceeded to teach his courses anyway and continued to do so until his death in 1687.[61] Marggraf's chemistry course followed the traditional didactic structure: a brief theoretical introduction followed by recipes and operations. Stemming from his university education, he espoused a traditional conception of chemistry as pharmacy, limiting his preparations to pharmaceutical recipes and common reagents needed for such recipes. Marggraf also taught private courses on *praxis medica*, so his chemistry course, following the established university model, could be seen as a supplement to his medical teaching. At the start of his chemistry course, Marggraf spent more time than De Maets in explaining the *theoria* of chemistry. He advocated five chemical principles (salt, sulfur, mercury, water, and earth) as the foundation for matter and included a discussion of chemical "instruments," which included standard laboratory equipment—furnaces and vessels—as well as "philosophical" instruments, such as *menstrua* (i.e., solvents) and fire (actual fire, not the peripatetic element). Marggraf paid greater attention to structure than did De Maets, dividing his course into four introductory "chapters" in which he covered theoretical matters. Each of these chapters was subdivided into numbered theses, which focused on definitions and explanations of terms and concepts. The fifth and, by far, largest chapter presented chemical operations and a long list of pharmaceutical recipes and their uses.[62]

A second lecturer, Jacob le Mort (1650–1718), who was the son of an apothecary, studied theology briefly in Leiden before apprenticing himself to Johann Glauber in Amsterdam and then to De Maets himself in Leiden. In 1675 Le Mort passed the Leiden apothecaries' examination, opened a chemist's shop, and began to offer private chemical courses. In 1677, he was fined by the Leiden Surgeon's Guild for calling on patients, so he opted to obtain a medical degree from Utrecht in 1678.[63] Le Mort's chemical lessons differed from both De Maets's and Marggraf's in that he grounded his chemical theory in Cartesian matter theory, preferring to explain chemical change in terms of the motion (*motus*) and rest (*quies*) of particles. Le Mort held that the traditional chemical principles were reducible to two "physical" principles: fluidity (*fluidis*) and constancy (*firmis*). Fluidity, or the principle of motion, was caused by an active, subtle matter, whereas constancy, or the principle of rest, was a product of the weight of (normal) matter. Le Mort also invoked the standard Cartesian tropes of particle shape, porosity, and texture to explain the selective interactions of chemical species. Like his contemporaries, however, the bulk of his lectures focused on chemical recipes both pharmaceutical and otherwise.[64] Le Mort also taught a course strictly on pharmacy, which focused solely on remedies and their uses, omitting most of the theoretical introduction found in his chemistry course.[65]

In this environment of three competing lecturers, the Leiden students were free to choose, arrange, and integrate material from each course in a variety of ways. Students interested in chemistry often attended the chemical courses of two or even all three lecturers, and copies of lecture notes taken from the chemists' courses circulated widely among the medical student population. Often, students extracted interesting recipes or ideas from individual courses and disregarded the rest, as shown by the circulation of *excerpta* manuscripts from each course.[66] Inevitably, a set of notes taken from the courses of all three chemists was edited and published. The resulting book, *Collectanea Chymica Leydensia* (1684), appeared in Leiden under the name of an English student, Christopher Love Morley (Leiden M.D., 1679). The book named De Maets, Marggraf, and Le Mort in its subtitle but made little effort to distinguish between the programs of each individual chemist. The *Collectanea* mimicked the traditional textbook structure by first presenting brief theoretical introductions, cribbed from both Marggraf's and Le Mort's courses (but significantly not from De Maets's), offered as "prologues" to the main text. The body of the book presented the collected preparations from all three courses arranged according to the alphabetical order of their products. The chemist from whose course each recipe came was identified in the margin.[67] In effect, the *Collectanea* was a reflection of chemistry as interpreted by students in Leiden during this period. The book allowed its reader to study, evaluate, and combine ideas and recipes (sometimes from three or more alternatives) as he saw fit, which was exactly the manner in which students in Leiden absorbed chemical instruction. Largely because of this versatility, the *Collectanea* went through five editions by 1702.[68]

All three chemists condemned the book. Le Mort and Marggraf blamed the errors that they perceived in the book on one another, igniting a war of words conducted in print.[69] De Maets responded with his own *Prodromus Chemiae Rationalis* (1684), a long pamphlet defending the merits of his course and especially its inclusion in the Leiden medical curriculum. In the introduction of the *Prodromus*, he too complained that the *Collectanea* was riddled with mistakes and railed that the printer had committed an act of piracy, asserting that "these thefts and crimes are intolerable to learned men."[70] As for the book's factual errors, he suggested that they were introduced through the process of copying and compiling the various sets of notes, and he provided corrections for several apparently egregious mistakes.[71] The most upsetting aspect of this affair for De Maets, however, was the manner in which the *Collectanea* depicted him and his chemistry. An introductory letter from the "typographus" of the *Collectanea* described De Maets as a "neomagus," who, among other things, brought chemical "arcana" to public light. Although this characterization was probably meant to compliment De Maets, he sternly rejected this association with the occult arts. He had, in fact, worked to dissociate his chemistry from this perception since it alluded to a part of chemistry that was increasingly ridiculed in academic circles.[72] In the *Prodromus* he asked rhetorically how he could be appointed professor if this characterization were true, suggesting that the description was a "crass error and rude argument."[73]

Nevertheless, the structure of the *Collectanea* and the lack of a theoretical introduction from De Maets implied that his chemistry had no theoretical foundation and little pedagogical structure. In response, De Maets attempted to elaborate on the philosophical foundation for his approach to chemistry. He described his method of demonstrating recipes and experiments as a process of "uncovering the secrets of nature," following in the footsteps of the Florentine Academy and Francis Bacon: "Neither contriving nor fashioning, but [rather] discovering what nature produces or may bear."[74] This Baconian method, De Maets asserted, was the same method followed by anatomists, in which one collected observations (*data*) on the structures of the body (*machina*) and, from these data, constructed laws in accordance with its workings. Chemistry assisted in this endeavor by examining such things as the chemical composition of body parts and the "fermentations" that take place in the liver, pancreas, and other glands. In following the method of the anatomists, he argued, one must avoid making improper, theoretical assumptions: "feigning dubious, obscure and, indeed, false hypotheses, one may fall into a thousand absurdities by no means guided by the structure of our bodies or in harmony with its motions."[75] De Maets contended that chemistry was not overtly theoretical because premature theorizing led to error. Ultimately, he maintained that chemistry deserved its place in the medical faculty because it both aided in the study and treatment of the body and shared the same empirical philosophy as the other medical fields.

His nod to anatomy and physiology may have justified De Maets's position on the Leiden medical faculty in the short term, but it did not secure the institutional position of chemistry. When De Maets died unexpectedly in January 1690, Jacob le Mort petitioned the Leiden curators for permission to give university-sanctioned lectures on chemistry, probably as a stepping stone toward succeeding De Maets as professor. The medical faculty, however, moved to block Le Mort's request. Medical professor Anton Nuck (1650–1692) composed a memorandum to the curators which argued that such private lectures reduced attendance at the professors' lectures, thereby tarnishing the "luster" of the university.[76] Nuck's memorandum was motivated at least in part by Le Mort's advocacy of Cartesian philosophy, whose influence the professors on the medical faculty were trying to expunge. As a result the curators followed the same cautious approach that they took with De Maets. They named Le Mort *praefectus* of the chemical laboratory, but he was not permitted to announce his lectures at the university. Unlike De Maets, he was not given an allowance for fuel or equipment.[77] Le Mort was eventually allowed to announce his chemistry lectures the following November and, as the result of a shortage of medical professors after the untimely death of Nuck in 1692, he was allowed to lecture on medicine as well.

Le Mort at first worked diligently toward obtaining an ordinary appointment on the medical faculty. One strategy he used was to argue that his approach to chemistry was theoretically grounded. In 1696 he published a 150-page theoretical and rhetorical opus, *Chymia Verae Nobilitas et Utilitas*, which explained his theoretical views on chemistry by employing concepts from Cartesian natural philosophy. In his dedication to the Leiden curators, he extended the arguments of De Maets by stating that chemistry had been cultivated even in antiquity as a part of medicine, and so it was right and necessary to cultivate it now.[78] By early 1697 the curators were ready to offer Le Mort a chair in chemistry on the medical faculty. Dutch stadholder (and king of England) William of Orange blocked their selection, however, because he and his orthodox, conservative Dutch supporters objected to the appointment of Cartesians to the Leiden faculty.[79] The curators were not able to appoint Le Mort to the medical faculty until William's death in 1702. In the meantime an increasingly disgruntled Le Mort felt no need to teach regularly, and during two winter terms (1697, 1700) he did not teach at all.[80]

From the perspective of the medical students as consumers of knowledge, the period from 1670 to 1690 was the golden age of Leiden chemistry. Both students and lecturers reinterpreted and rearranged the field as they saw fit with little interference from university authorities. The result was a *chemia Leydensiensis* that was a mix of approaches and operations deriving from craft traditions, chemical medicine, alchemy, and natural philosophy, synthesized and shaped to fit the traditional, didactic chemical structure. Despite this era of productive exchange, chemistry remained a marginal, low-status enterprise at the university. The Leiden curators and the medical faculty could have intervened in support of Carel de Maets, the professor of chemistry, and suppressed the competing lecturers and their approaches. Yet they allowed the private lecturers to proceed because chemistry was not a central concern for the university, and the competitive atmosphere among the chemists attracted more interest from students than De Maets's courses could alone. After the deaths of Christiaan Marggraf and De Maets, chemistry at Leiden experienced a marked decline. Jacob le Mort's lectures were irregular, and no other competing lecturer emerged during the 1690s to fill the gap or compel Le Mort to improve.

BOERHAAVE'S CHEMICAL EDUCATION

When Herman Boerhaave began to study chemistry in 1691 or 1692, courses on the subject were not readily available in Leiden. De Maets had died two years earlier, and Le Mort had not yet established his course at the university. As Leiden students had done earlier in the century, Boerhaave had to create his own plan of study relying on whatever local resources he could find. As he described this period of his life in his "Commentariolus," "he [Boerhaave] cultivated chemistry day and night."[81] The resources that he used to undertake this study included the books at the University of Leiden's library, which held the newly acquired Vossian collection. Just as he had read the university library's holdings on medicine from Hippocrates to Sydenham, when he turned his focus on chemistry he seems to have read every work he found in the library's extensive collection. In addition, manuscripts of the didactic courses of De Maets, Marggraf, and Le Mort still circulated. Boerhaave obtained copies of De Maets's and Marggraf's courses taken by an earlier student, Johannes Broen.[82] Combined with his ongoing study of medicine, this eclectic, yet highly literary training shaped his approach to chemistry. Boerhaave developed the practical, laboratory skills of an artisan chemist, but his interest in chemistry stemmed from a desire to investigate knowledge claims in chemistry and medicine.

To gain practical training in chemistry, Boerhaave had to find a suitable mentor. As he recounted much later, he worked with his brother on chemical experiments conducted at their home, although he did not record the exact nature of these experiments.[83] Around 1692 or 1693, he made an arrangement with a local apothecary, David Stam (1633–1711), to receive practical training in Stam's laboratory. Stam himself had worked in the private laboratory of Franciscus Sylvius and, if his own statements are to be believed, he received a medical degree as well, although no record of this degree exists.[84] Nevertheless, Boerhaave thought very highly of Stam, whose service to Boerhaave was reported by the professor's former student and biographer William Burton: "great is the advantage, which a skillful and ingenious operator . . . may afford a student in this branch of philosophy; and perhaps much greater, than in years he could obtain from Books, and his own labours."[85] Boerhaave did not record what he did or learned in Stam's laboratory, but likely he became familiar with common chemical equipment and operations that were part of any apothecary's usual work.

In addition to practical training, Boerhaave also received a close look at the debates between rival chemical schools. Stam prepared his only published work, an edition of John Francis Vigani's *Medulla Chymiae* (1693), while Boerhaave was an assistant in his laboratory. The *Medulla* was a short primer on "recent" developments in chemistry, modeled on the structure of traditional, didactic textbooks. The book began with a definition and etymology for *chymia*, briefly examined the chemical principles, and then proceeded to discuss the properties and operations of chemical species, arranged in a somewhat idiosyncratic order: salts, tinctures, plant extracts, animal extracts, and metals.[86] Vigani did not intend the *Medulla* to be a comprehensive textbook but rather an attempt to place Robert Boyle's chemical philosophy into a didactic form. Following Boyle, Vigani rejected the various systems of chemical principles, including the Paracelsian *tria prima* (salt, sulfur, and mercury), the peripatetic elements, and the acid and alkali theory of Stam's mentor, Sylvius. Instead, Vigani posited that chemical change should be explained in terms of "atoms or corpuscles," citing Boyle's *Sceptical Chymist* as the authoritative source for this claim.[87] The rest of the book attempted to place a few traditional and exotic (e.g., "viper stones" and "Cantharides") preparations within this Boylean theoretical framework, although the new theoretical perspective seemed to have little impact on the descriptions of the operations themselves.

By contrast, Stam's edition of the *Medulla* was a defense of traditional chemical principles and practices in response to Vigani's critique. In a preface to the book, Stam affirmed his commitment to the *tria prima* as the "most clear vocabulary of nature" and asserted their centrality to chemical investigation.[88] He kept Vigani's text largely the same, but he added long footnotes criticizing claims with which he did not agree. In the section where Vigani advocated Boyle's atoms, Stam defended the work of his mentor, Sylvius, in a long footnote—longer than Vigani's original argument—and posited that, unlike atoms, the *tria prima* were empirically demonstrated and universally accepted. To counter Vigani's citations of Boyle, Stam generated his own list of traditional chemical authorities, such as Joan Baptista van Helmont and Basil Valentine, to support his position.[89]

Ultimately, Boerhaave did not support Stam's defense of traditional chemistry. Although he appreciated artisan chemists for their practical skill, Boerhaave became a partisan of Boyle's skeptical program in which knowledge claims were tested through experimentation. He openly acknowledged his intellectual debt to Boyle in all of his published writings on chemistry. In his textbook *Elementa Chemiae* (1732), he called Boyle "the great Master" and stated that he grounded his own program on Boyle's principles.[90] Boerhaave, however, had equal respect for Van Helmont, one of Stam's favorites. He read Van Helmont's main work, *Ortus Medicinae* (1648), several times during his life, committing large parts of the text to memory.[91] In effect, he did not see these two authors as opponents but rather as complementary sources of chemical knowledge and practice. Both published accounts of their chemical experiments, which they pursued to investigate knowledge claims rather than work out chemical recipes.[92] This form of chemical practice, which focused on investigating the natural philosophy of chemistry rather than on mastering its recipes and commercial products per se, appealed to Boerhaave.

Boerhaave's focus on theoretical questions in chemistry can be gleaned by looking more closely at the texts he studied during his literary education in chemistry at the Leiden university library. Much of what we can deduce about his early study derives from over three hundred folio pages of extant reading notes, probably dating from this early period in his career in chemistry.[93] The content of these notes indicates Boerhaave's primary interests at the time: the analysis, synthesis, and transmutation of metals and the preparation of metallic medicines. The notes from Boyle's work, including the *Sceptical Chymist* (1661) and *Some Considerations Touching the Usefulness of Experimental Natural Philosophy* (1663), all refer to passages that discussed the properties of mercury or antimony, metals of alchemical and medicinal interest to both Boyle and Boerhaave.[94] In another set of notes titled simply "mercurius," Boerhaave recorded passages from Van Helmont's *Ortus*, which described the composition of quicksilver as the union of "mercury" and two types of "sulfur." As Boerhaave recorded, once freed from these "sulfurs," the pure "mercury of mercury" could reduce other metals to their prime matter.[95] He believed that this mercury was Van Helmont's *alkahest* or universal solvent and was identical to the sophic mercury of the mercurialist school of alchemy (described at the start of this chapter).[96] These notes, reflecting Boerhaave's philosophical interests, do not contain recipes or procedures but rather examine chemical theories and the special properties of interesting chemical species.

In the passages that Boerhaave recorded, both Boyle and Van Helmont elaborated on the mercury/sulfur dyad theory of metals. This theory held that metals were composed of a material, fluid "mercury," which was "fixed" by a coagulating agent, "sulfur." The various known metals were distinguished from one another based on the level of impurities present in the metal. Gold was made from the purest mercury and sulfur, whereas lesser metals were contaminated with "gross" or "vulgar" sulfurs, earths or "arsenick."[97] This theory probably originated in Arabic alchemy and became popular in the West through the influence of the *Summa Perfectionis* of "Geber," a pseudonym of the fourteenth-century monk Paul of Taranto.[98] Boerhaave also took extensive reading notes on the *Summa Perfectionis* as well as on dozens of other texts that discussed the dyad theory and its alchemical applications.[99] In effect, his chemical education included a great deal of book research in the chemical literature in order to map out the various versions of the dyad theory and work out their consequences.

After ending his university studies in 1693, Boerhaave expanded his understanding of chemical theory by conducting his own experiments designed to examine the nature and properties of metals. He conducted these experiments in a small laboratory that he maintained in a shed next to his house in Leiden. One of his early projects was an attempt to make the sophic mercury of Eirenaeus Philalethes. After conducting the required literary work to decipher Philalethes's recipe and performing some preliminary experimentation, Boerhaave began a first attempt at sophic mercury on June 1, 1697. He crafted an amalgam of quicksilver and *regulus* of antimony reduced by iron (*regulus antimonii martialis*) with the help of a bit of silver acting as a flux. He "digested" this amalgam for four days and washed away the resulting "blackness" with water. Finally, he sublimated the resulting white amalgam, recovering the original mercury. In his experimental notes, Boerhaave called this mercury the "first eagle" (*primus aquilae*), following Philalethes's terminology.[100] After a hiatus until December, he began the process of further purifying his mercury with more antimony and sliver amalgams, until he had taken the "twelfth eagle" by March 1698.[101] Unfortunately, Boerhaave did not record the properties of his resultant mercury in his experimental notes. Although this lack of comment suggests that his results were inclusive, he was not dissuaded, since he made another attempt at sophic mercury in the 1720s.

In another line of inquiry, Boerhaave sought to examine the chemistry of antimony, a project he began immediately after his attempt at sophic mercury and, perhaps, derived from it. Antimony was a popular yet controversial topic among chemical practitioners because of its interesting chemical properties and use as a powerful purgative in medicine. In the seventeenth century, the term "antimony" referred to stibnite, modern antimony sulfide ore. Assayers had long known of antimony's use as a reducing agent.

Roasting ground antimony with iron filings produced a sulfurous iron calx and metallic antimony, called "*regulus* of antimony."[102] To chemists who followed the mercury/sulfur dyad theory, this operation was interpreted as the crude antimony absorbing the iron's fixing or metallic sulfur, transforming (i.e., reducing) the iron into a nonmetal: mercury loosely combined with crude sulfurous impurities.[103] The power of antimony to absorb fixing sulfurs suggested to some chemists that antimony could be a powerful tool in both alchemical operations and medicine. Antimony and antimonial remedies became the icons of the new, "chemical" medicine during the late sixteenth century and were seen as a challenge to the traditional, herbal pharmacy of the Galenists.[104] Meanwhile, alchemical writers such as Alexander von Suchten and George Starkey, via Philalethes, used antimony as a purifying agent in their attempts to make sophic mercury.[105] The interest in antimony had not died down by the end of the century, as shown by Nicolas Lemery, who, while Boerhaave was conducting his work on antimony, began his own comprehensive study of the chemistry of antimony at the Académie royale des sciences in Paris.[106]

Boerhaave conducted experiments on crude "antimony," its sulfur, and the *regulus* of antimony. In one set of experiments, Boerhaave examined the sulfur generated by the production of *regulus* of antimony through various methods. Several chemical authorities, from Van Helmont to Angelo Sala (1576–1637), claimed that the "sulfur of antimony," properly prepared, exhibited wonderful medicinal properties.[107] Boerhaave had taken extensive reading notes from the work of Boyle and Otto Tachenius on the chemistry of antimony and on the methods to obtain its sulfur from crude antimony by various means.[108] From these notes, he tried various techniques of removing antimony's sulfur, most of which involved displacing the sulfur with *aqua regia* (a mixture of nitric and hydrochloric acids that could dissolve gold). He then investigated the properties of this sulfur by using it to make a pharmaceutical preparation called tincture of antimony, which was a strong emetic. The question for Boerhaave was whether this medicinal property derived from a subtle "arsenic" that was part of the antimony.[109] In a related experiment taken from Van Helmont's *Ortus*, Boerhaave dissolved cinnabar of antimony (mercury combined with the sulfur obtained from stibnite) in wine to produce another medicament, which was supposed to cause "wonderful effects."[110] By June 1702, this research had come to an end. Boerhaave declared in his notes that the scoriae of antimony was "pure vulgar sulfur" and thus had no special virtues beyond those of common sulfur.[111] This project reflected the nature of Boerhaave's chemical work—

determining the composition and properties of important chemical species and evaluating the claims made about them in the chemical literature.

CONCLUSION

Boerhaave found that most of his experiments on metals and metallic medicines did not seem to support the claims made about them. Although unfruitful, this work reflected his theoretical interests in chemistry. He did not spend a great deal of his time studying and perfecting mundane chemical operations but rather focused on processes that could expand his understanding of controversial or philosophical questions. He did not, however, ignore or reject the artisan chemistry of the didactic tradition. Knowledge of practical techniques and recipes formed the foundation of every chemist's training, regardless of their later interests and practices. In his textbook *Elementa Chemiae*, designed for university medical students, Boerhaave recommended several didactic textbooks as primers in chemistry, including Beguin's *Tyrocinium Chymicum* and Lemery's *Cours de chymie*.[112] The didactic tradition did not, however, teach the chemical novice how to *investigate* the pressing chemical questions of the day, such as the composition of metals and the causes of medicinal "virtues," that concerned Boerhaave and other educated physicians and philosophers. For this type of chemical practice, he looked to the writings of alchemists, chemical physicians, and natural philosophers, and also to his own experimental work. When Boerhaave became a lecturer in chemistry at Leiden, he endeavored to create a new kind of chemistry course by marrying these two traditions.

CHAPTER THREE

The Institutes of Chemistry

In January 1702, the Leiden curators and burgemeesters granted Herman Boerhaave permission to teach chemistry and anatomy at "the earnest request of some foreign students."[1] He had been lecturing in the *institutiones medicae* course in the medical faculty for eight months and had been conducting chemical study and experimentation at his residence for several years. The curators' decree would allow him to bring his chemical expertise into the medical faculty by giving private lectures sanctioned by the university. The primary challenge for Boerhaave was that he had neither taught nor even attended a chemistry course. He had models for his course—the collection of didactic textbooks in the University of Leiden library and sets of lecture notes from previous lecturers—but he rejected the pedagogical structure and theoretical framework of the traditional chemistry course. He opted instead to create a new course that could be better integrated into the theoretical and pedagogical norms of the medical curriculum at Leiden. In effect, Boerhaave strove to structure his new chemistry course more like a medical course.

The main way in which Boerhaave made this change was by shifting the pedagogical aim of his chemistry. Traditional didactic chemistry tended to focus on teaching students, through the use of recipes, how to make products that an artisan chemist should know how to make: medicaments and other commercial substances. Although this knowledge was, indeed, useful and practical, this type of knowledge was not the sort that medical students traditionally mastered in a university medical faculty. University courses focused on the rigorous definition of terms and concepts, theoretical principles and arguments, and the rules (or precepts) of medical practice. Medical students mastered this type of largely theoretical knowledge, *scientia* in its classical definition, so that they could exhibit their erudition in oral disputation exercises and examinations, which were the main ways they were evaluated by their professors. To make chemistry fit into the medical faculty, Boerhaave had to design a course that focused on theoretical principles and precepts and the definition and taxonomy of chemical entities, not on practical recipes. Chemical knowledge presented in this fashion could be integrated into the pedagogical practices of the medical faculty at Leiden and, ultimately, deployed to help explain medical problems and treatments.

Boerhaave devised a *scientia* for chemistry that was much more rigorous than those found in the short theory sections of most didactic textbooks. He grounded his chemistry in the same empirical natural philosophy that he advocated for his medicine. This approach (outlined in chap. 1) was experimental in its methodology and assumed a mechanical world composed of material corpuscles in motion. Following arguments by Robert Boyle and others, Boerhaave rejected the various systems of chemical principles found in didactic textbooks. According to Boyle, the entities that chemists presented as chemical principles were, in fact, artifacts of the operations that were used to analyze them.[2] Instead, Boerhaave developed a theoretical approach to chemistry that was designed to understand the effects of chemical operations on matter. Building on the work of Johannes Bohn, a professor of medicine from Leipzig, Boerhaave argued that chemists manipulated matter by understanding how its chemical properties interacted with the chemical "instruments." These "instruments"—fire, air, water, earth, and chemical menstrua—were the natural tools that the chemist used to conduct his operations, and that the natural philosopher studied in order to understand chemical change. The instrument theory became the theoretical core of Boerhaave's chemistry courses. Because Bohn had already presented a version of the theory in an academic context (as a series of dissertations), Boerhaave found that he could adapt Bohn's approach to the pedagogical context of the Leiden medical faculty.[3]

This chapter presents a contextual reading of Boerhaave's first chemical courses, which he offered in 1702–3, as presented in his own manuscript notes. These courses include a private lecture course on chemistry, which he first taught starting in January 1702 and again in October of that year, and a course on chemical operations, which included demonstrations of those operations, in early 1703. An examination of both courses shows how Boerhaave applied the pedagogical method from his medical courses to chemistry and also reshaped the conceptual foundation of didactic chemistry to make it fit the academic notion of an "art." Ultimately, Boerhaave's reworking of didactic chemistry to fit the norms of the Leiden medical faculty fundamentally changed not only the status of chemistry at the university, but also its pedagogical structure, theoretical framework, and place in the medical curriculum.

CHEMISTRY INTO MEDICINE

In May 1701, soon after he had been appointed as a lecturer on the medical faculty at Leiden, Boerhaave urged the Leiden university community to make chemistry a part of the medical curriculum. In his *Oratio de Commendando Studio Hippocratico,* he encouraged Leiden medical students to study chemistry, which had been "purged [of errors] by the diligence of the present age." To assist the student in this study, he recommended the works of Joan Baptista van Helmont, Robert Boyle, and Otto Tachenius.[4] Yet when Boerhaave was allowed to give private lectures on chemistry a few months later, he found that there were no extant chemistry texts that fit his needs for the medical curriculum. The reason that he chose to mention Van Helmont, Boyle, and Tachenius was for their "philosophical" approach to the chemical art, not necessarily for specific claims found in their works. For Boerhaave, the "philosophical" approach to chemistry meant the use of chemical operations to uncover the properties and powers of chemical species via experiment, as he had done in his private chemical work. Knowledge of chemical properties and the techniques of discovering those properties were what made the study of chemistry significant for medicine. This approach provided reliable knowledge regarding actions of medicinal substances and bodily fluids, which was eminently useful for later medical studies. In Boerhaave's mind, this was the approach to chemistry that Van Helmont, Boyle, and Tachenius followed, but none of these authors had penned a textbook.

Boerhaave's program for chemistry reflected the approach to knowledge found in the Leiden medical faculty. The Leiden medical curriculum, like that of most universities, was designed to produce gentlemen professionals.[5] The physicians produced through university education occupied the top rank on the medical hierarchy largely because their knowledge of the body and of medicine was theoretical and not derived solely from a practical skill. Despite the University of Leiden's reputation for a hands-on approach, the basis of pedagogy in the medical faculty remained the lecture and the primarily theoretical knowledge that the lectures conferred. Apart from the demonstrations that the student observed in botany and anatomy courses, the content of the two core medical courses in Leiden dealt solely with the elaboration and application of theoretical knowledge. The *institutiones medicae* course reviewed theories of physiology and principles of diagnosis, disease prevention, prognosis, and treatment. The *praxis medica* course examined specific diseases, their diagnosis and treatment ideally based on the theories and principles one learned in the *institutiones*. Medical degree candidates were usually examined on material from only these two courses. As the medical faculty statutes of 1576 stated, degree candidates had to pass two oral examinations. The first was on a specific Hippocratic aphorism that had "a universality of method." After receiving his aphorism, the student had one day to prepare an explanation of the medical principles asserted in it and to prepare answers to possible questions. The second exam was an hour-long disputation concerning the proper treatment of a specific disease.[6] The point of both of these examinations was to test whether the student had understood and could apply the theoretical principles of medicine presented in the core medical courses. During the examinations, the student was expected to do more than simply recite and support the received view of his advising professor. He also had to explain why differing views on the issue at hand were incorrect. Thus, the whole process was very academic: the candidate had to display his familiarity with a variety of positions and be able to argue effectively for the preferred one. Within this context, the practical concerns of medicine were of secondary importance to the theoretical.[7]

When a medical candidate graduated, he participated in a ceremonial disputation exercise. This formal public debate focused on the candidate's dissertation: a set of propositions that he had written on a specific medical topic. In traditional academic disputations, a *defendens* was required to defend with logical arguments a set of numbered propositions or theses on the specific topic at hand. Each thesis presented a simple and direct statement asserting one important definition, interpretation of fact, theoretical principle, or specific argument, although in practice the explanation or elaboration on the central point of the thesis could extend over several pages. The *promotor* functioned as the primary opponent of the *defendens* during the debate, questioning the validity of the theses. In theory, however, anyone in attendance could question the *defendens*, and many dissertations were printed and circulated beforehand to encourage the university community's participation.[8] In Leiden, the "inaugural" disputation, as part of the graduation ceremony, had the graduating medical student participate as the *defendens,* while his advising professor acted as *promotor.*[9]

In a pedagogical context in which students were examined orally in this manner, the practical goal of medical education was, in part, to equip students with the skills and knowledge base needed to perform well in debates over medical theory. Boerhaave and other medical professors therefore organized their courses, both in their notes and in their textbooks, as sets of theses. For example, the first edition of Boerhaave's medical textbook *Institutiones Medicae* (1708) was composed of 1,002 theses covering the range of medical theory and practice that he addressed in his lecture course of the same name. His *Aphorismi de Cognoscendis et Curandis Morbis* (1709), which derived from his *praxis medica* course, similarly contained 1,479 "aphorisms" on the treatment of disease.[10] Written as theses, the content of Boerhaave's lectures and textbooks could easily be employed to prepare students for academic exercises, where they were required to defend and expound upon such theses in public debate.

Medical lecturers, like Boerhaave, structured their courses according to a definite pedagogical order or method. They grouped related topics—physiological principles, types of diseases, signs of illness, and so forth—according to taxonomic categories based on theoretical principles. Like Libavius's ordering of the chemical art in his *Alchemia,* the method of ordering employed by a medical lecturer provided an intellectual structure for an otherwise chaotic morass of medical concepts, precepts, and practices. The method defined the theoretical topics and parameters of the lecturer's particular approach to medicine and also served as a mnemonic device for the medical student.[11] One traditional method in medicine simply ordered diseases based on the body part affected, starting with the head and ending with the feet. By the eighteenth century, medical lecturers often opted for more sophisticated systems of ordering that reflected their or their institution's particular approach to medical education. Boerhaave's *Institutiones Medicae,* for example, was divided into five general headings: the animal economy (i.e., basic physiology and the systems of the body), diseases, signs (of disease and health), the maintaining of health, and the curing of diseases. Each of these general headings was subdivided; for example, the general heading "diseases" was divided into two topical groups containing those affecting the solid and those affecting the fluid parts of the body.[12] For Boerhaave, this method reflected the medical topics that needed to be covered in the Leiden medical faculty's *institutiones* course as well as his own theoretical ideas regarding the structure of the body.

In 1702 Boerhaave devised a new method for ordering a chemistry course—a method that shifted the emphasis from practical recipes and operations to the theoretical principles underpinning those operations. In effect, he had to shape his chemistry according to a method that he could integrate with the method he used to structure his medical courses. His first task was to develop a theory course for chemistry. The first division in any academic treatment of an "art" according to a method differentiated between theory and practice. In all university subjects the theoretical principles of a field were covered first, followed by their practical applications, which were guided by the previously established theory. In the medical faculty, basic medical theory was covered in the *institutiones medicae* course, and the basic principles established in the *institutiones* course were applied to specific examples in the *praxis medica* course. In an example more applicable to Boerhaave's immediate situation, the Leiden physics (or natural philosophy) curriculum also followed this model: first, *physica theoretica* was taught as a lecture course; then *physica practica* was taught in the physics theater through experimental demonstrations. As we saw in chapter 2 with the chemistry course of Carel de Maets, the standard didactic chemistry course was already seen predominately as a form of *practica*, either for physics or for medicine as pharmacy. Yet Boerhaave's first chemistry course in the winter of 1702 was a lecture course with no demonstrations; it was *chemia theoretica*. The course of chemical operations, which he gave the following fall to augment his lectures, presented a pedagogically sophisticated version of the standard didactic "course," adding the *chemia practica* component to his chemistry curriculum. For his lecture course, however, he needed to develop a methodized *theoria* for chemistry that fit the medical model.

Boerhaave found a model for his chemistry course in a series of published dissertations written by the Leipzig medical professor Johannes Bohn (1640–1718). Bohn's *Dissertationes Chymico-Physicae* (1685) consisted of fifteen dissertations on "chemical" topics, each containing twenty or thirty numbered theses. Each thesis ranged from a few lines to an entire page in length and presented a definition or theoretical proposition relating to the nature of the chemical art, a "chemical" entity, or chemical operation. Bohn had written the theses for his students to defend publicly in the academic exercises conducted in the University of Leipzig's medical faculty. Just as in the Leiden exercises, the aim of the student *defendens* was to argue for the logical import of each thesis, thereby demonstrating that he understood the philosophical principles behind the thesis. As shaped by this context, Bohn's theses focused on definition, theoretical argument, and explanation rather than on chemical recipes or practice per se. For example, his first two dissertations on the dissolution and combination of bodies examined the two traditional categories of chemical operation as defined by the didactic tradition. Rather than describing specific operations, however, Bohn addressed the philosophical problem of the type and nature of the elements or principles into which the chemical art divided bodies. His next fours dissertations elaborated on the chemical "instruments": fire, air, water, earth, and chemical menstrua, which Boerhaave adopted as the core of his lecture course. These four dissertations presented theoretical explanations regarding how each of these "instruments" caused various types of motions, dissolutions, and coagulations of chemical species during chemical operations. The remaining dissertations addressed specific types of chemical operations—digestion, fermentation, sublimation, calcination, and so forth—but again, Bohn focused on defining terms and establishing theoretical explanations for phenomena and not descriptions of procedures.[13]

Boerhaave emulated the structure of Bohn's *Dissertationes Chymico-Physicae* when he put together his first chemical lecture course. In preparation for assembling his course, he composed an outline from Bohn's book in which he recorded the theses that he later incorporated into his lecture course. He organized the outline into eight "chapters." The first two related to the history of chemistry, which Boerhaave cribbed from the preface to Bohn's book and later augmented greatly with material from other sources. The third chapter addressed the formal definition of chemistry, and the last five collated theses regarding Bohn's notion of chemical "instruments." Like Bohn's dissertations, Boerhaave's chapters were organized as a series of numbered theses, each consisting of a definition or statement of theory often cribbed directly from Bohn. The outline, however, was not an exact representation of Bohn's book. Boerhaave excluded material with which he apparently did not agree and did not take any notes from Bohn's dissertations on specific types of operations.[14] Clearly, what interested Boerhaave was Bohn's definition of chemistry and the theory of the chemical instruments. When Boerhaave drafted his course, he wrote it as a series of 264 theses that generally followed the same order as his outline of Bohn's *Dissertationes*.[15]

Boerhaave structured the course through a rigorous method of definition and division. After beginning with a brief history of chemistry, he initiated the main part of his lecture course by stating a definition of chemistry (*chemia*).[16] All methodized academic courses, and even didactic chemical manuals, ideally began with an all-encompassing definition, from which all the topics of the art were extracted through subsequent division and definition.[17] Boerhaave defined chemistry as follows: "Chemistry is the art that teaches to employ operations, by which bodies, that are able to be enclosed or held in vessels, are broken apart or composed by the power of certain instruments so that the effects and mutations of bodies may be discovered, truly efficacious, safe, and acceptable medicines may be discovered, and various arts may be properly treated."[18] Boerhaave's definition conveyed to the student the proper methodology and function of the chemical art: the experimental use of instruments to further the progress of natural philosophy, medicine, and the "various arts." In a later thesis, he expounded upon this definition and explained how it framed the pedagogical structure of his course. He asserted that all compositions and separations in chemistry may be reduced to the consideration of "objects" (or bodies, i.e., chemical species) and "actions" (or operations). Boerhaave differentiated between the two terms based on epistemology: the "objects" were known "mathematically" through their figure, cohesion with other bodies, and motions, while the "actions" or "operations" were only disclosed experimentally through induction. He then defined "instruments" as the media through which the chemist effected his operations on objects.[19] These tedious definitions of simple terms and the explication of their relationship to one another may have seemed philosophically trite, but they served an important pedagogical function. "Objects," "instruments," and "operations" were the organizational categories with which Boerhaave structured the remainder of his course. Through the rigorous explication of these categories, he indicated both how his course fit together conceptually and how that conceptual structure related to his aim of discovering the "effects and mutations" of bodies, finding effective medicines, and improving the various arts.

Boerhaave addressed each of the three pedagogical categories in order, beginning with the "objects" of chemistry. This section of the course described species of chemical simples (or reagents) arranged in a taxonomic order according to common natural history categories. In presenting this taxonomy, he maintained the method of definition and division. He divided the "objects" first into three kingdoms (mineral, vegetable, and animal) and then into various genera. For example, the mineral kingdom was divided into metals, semi-metals, salts, stones, and earths. He then defined the taxonomic characteristics of each of these genera and identified their most common chemical species.[20] Boerhaave's taxonomical divisions still derived from conceptual categories found in the didactic tradition. In the better and most recent textbooks, such as Nicolas Lemery's *Cours de chymie*, the de-

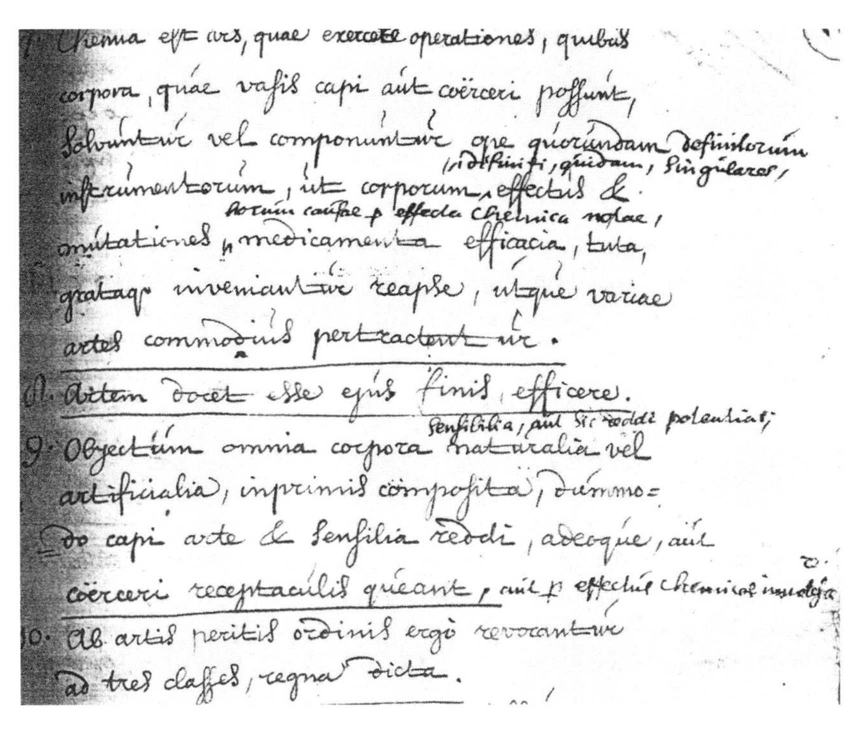
7. Chemia est ars, quae exercet operationes, quibus
corpora, quae vasis capi aut coërceri possunt,
solvuntur vel componuntur ope quorundam definitorum
instrumentorum, ut corporum effectus &
mutationes, medicamenta efficacia, tuta,
grataque inveniantur reapse, utque variae
artes commodius pertractentur.
8. Artem docet esse ejus finis, efficere.
9. Objectum omnia corpora naturalia vel
artificialia, inprimis composita, dummo=
do capi arte & sensilia reddi, adeoque, aut
coërceri receptaculis queant,
10. Ab artis peritis ordinis ergo revocantur
ad tres classes, regna dicta.

FIGURE 2. Theses from Boerhaave's *theoria* course. Source: VMA, MS 3, 15r.

scriptions of chemical operations were organized according to natural history classes, which were similar to the ones that Boerhaave utilized.[21] In this section of his course, there were no recipes or discussion of operations. Rather, he provided the definitions and descriptions of the various chemical species—their properties, where they could be found, and their uses—to which he could then refer again later in the course.

This orderly presentation of chemistry in terms of categories, definitions, and uses of chemical species constituted, according to the standards of the university, a theoretical presentation. Within the context of the medical faculty, Boerhaave's course presented material in a way that the student could integrate into his preparation for scholarly examinations and disputations. If, for example, the student was asked why a specific medicament was used to treat a specific disease, he could elaborate on the virtues of the medicament as a chemical species: what class of substance it was, what its chemical properties were, and how it might act on other substances in the body to effect a cure. By stating the reasons for his choice of medicament, he demonstrated his erudition in medicine, which was the mark of the educated, gentleman physician that the Leiden medical faculty strove to produce. Boerhaave structured the rest of his course, focusing on the chemical instruments and chemical operations, following this same method. For the chemical instruments he addressed each instrument in turn—defining it, discussing its properties, and suggesting its uses as a tool for chemistry. For the operations he defined the general categories of operations, explained how they worked, and reviewed their uses. How this method shaped these two topics is examined in the next two sections of this chapter.

THE INSTRUMENTS OF CHEMISTRY

In addition to providing a formal academic structure, Johannes Bohn introduced Boerhaave to the instrument theory of chemistry. The instrument theory was an alternative approach to understanding chemical operations, which was born out of a rejection of the traditional system of chemical principles (salt, sulfur, mercury, and sometimes water or phlegm and earth). As such, the instruments (fire, air, water, earth, menstrua, and chemical vessels) were the tools (natural and artificial) that the chemist manipulated during chemical operations. For Boerhaave, the instruments represented a theoretical and organizational approach to chemistry that was very different from that of the didactic tradition. The chemical principles as presented in didactic courses imbued substances in which they were mixed with specific properties, suggesting that chemical properties could be understood in terms of the composition of principles in bodies. By the turn of the eighteenth century, many chemists trained in the didactic tradition, such as Lemery or Boerhaave's mentor David Stam, argued that they could extract the principles of a body in physical form through chemical analysis.[22] By contrast, the instrument theory shifted the focus of chemical theory to the latent properties of chemical species, which were seen as inherent to individual species and were revealed only through interaction with the instruments. Ultimately, the instrument approach problematized exactly what the didactic tradition took for granted: the effects generated and properties revealed through the chemist's manipulations of matter.

The instruments were a heterogeneous group, both ontologically and functionally. "Fire" was an imponderable, particulate fluid, which acted as the cause of phenomena relating to heat, such as the expansion of a body's volume, flame, and the destructive dissolution of bodies. "Air" was a ponderable fluid, which exerted pressure on bodies necessary, for example, during combustion, and also acted as a medium to contain vapors and "spirits." "Water" was both a ponderable fluid and the most common solvent found in the chemist's laboratory. "Earth" was a simple, hard, insoluble body, which remained fixed in the fire, and whose function as an instrument was to "hold" volatile spirits, salts, and oils. Finally, the term "menstruum" did not refer to any specific species of body but generally to the power of any body to act as a solvent. Boerhaave's conception of a menstruum, however, extended far beyond the modern notion of a solvent. For example, he included as menstrua precipitations of salts in solution that modern chemists would identify as displacement reactions and later eighteenth-century chemists would call examples of "elective affinity." Similarly, he included in this category the "power" of fixed alkali (soda or potash) to absorb water from the air, deliquescing (i.e., melting) if left unattended.[23] Within the logic of Boerhaave's chemistry, the instruments functioned as the tools, which the chemist manipulated during his operations. This property was the only one that unified the instruments as a class of chemical object.

Boerhaave's adoption of the instrument theory came as a result of his rejection of the various systems of chemical principles, which was based on a critique of these entities that had developed during the second half of the seventeenth century. Chemical skeptics, most notably Robert Boyle, contended that the substances, which the "chymists" identified as principles or elements, were in fact compound bodies generated by the chemists' operations themselves. In other words, they were artifacts of the chemist's fire.[24] Boerhaave adopted Boyle's skeptical attitude toward traditional chemical entities, especially his critique of alleged, empirical demonstrations of the chemical principles, found in texts such as *The Sceptical Chymist* (1661), *The Producibleness of Chymical Principles* (first published as an appendix to the 1680 edition of the *Sceptical Chymist*), and the *Mechanical Origine or Production of Qualities* (1675). Boerhaave cited each of these texts in his earliest chemistry lectures and later in his *Elementa Chemiae*.[25] In these works, Boyle deployed experiments systematically to examine chemical claims in order to refute those that could not withstand empirical scrutiny. In this regard, Boyle's skeptical approach to chemistry was a useful tool within the context of Boerhaave's program to reform chemistry by suggesting a way to "purge the errors" from the field.

Taking his cue from Boyle, Boerhaave treated the chemical principles in the same way that he treated questionable claims by Cartesian physicians—as theoretical speculations. In his first chemistry course, he began his discussion of the chemical instruments with a thesis that asserted his Boylean skepticism: "The products of separation [i.e., chemical analysis] can neither be resolved into similar parts or elements, nor are they always produced by the separation alone without the mutation of their parts."[26] The lecture notes do not indicate how Boerhaave may have elaborated on and interpreted this statement in his lecture for his students. He did, however, clarify his position in much greater detail later in his *Elementa Chemiae*. Here, he argued that in chemical operations the same "powers" that disunite bodies "may produce in them likewise a great alteration, and we shall fall into error if we suppose that the [original] compound bodies in reality do contain these very elements."[27] The chemist's fire was just as likely to rearrange as to separate the "corpuscles" of compound matter and thereby produce new effects "which never [presented] themselves by any effect in the bodies while they were entire." Thus, he concluded, the "chemists" are mistaken when they claim to produce the "first elements of bodies" and "think they can determine the nature of compounds" from these elements. Indeed, the only entities that Boerhaave argued may rightly be called "elements" are the "corpuscles" or "atoms" that compose every body. But, he asserted, as a practical matter these corpuscles can never be collected in a pure state by the chemical art.[28]

Although Boyle's skeptical approach to traditional chemistry proved a valuable methodological resource for Boerhaave's reform of chemistry, it did not provide an alternative theoretical framework to the chemical principles. He adopted and adapted Johannes Bohn's interpretation of the instrument theory to construct a new framework. This process was made easier by the fact that Boerhaave saw Bohn as a philosophical ally. Bohn was a mechanist who depicted chemical phenomena in terms of matter and motion, but he did not endorse the "speculative" methods of the Cartesians.[29] He too was a great admirer of Robert Boyle and deployed Boyle's arguments in his *Dissertationes* to argue against the existence of chemical principles. In his first dissertation on the dissolution of bodies, Bohn addressed the question of whether one can resolve a "mixt" into its component principles or elements. He reviewed the examples upon which such assertions were made, which included various operations to extract the "mercury" of metals and to distill the "principles" of plants from crude plant matter. Bohn concluded that one could not "analyze" bodies into their elements. For this reason, he asserted that his chemistry would avoid (*eludunt*) the claims of the "philosophers of fire" (chemists) and "peripatetics" (Aristotelians) regarding the elements. In support of this position, he cited that "Illustrissimus Philosophus Anglicae," Robert Boyle, from his *Sceptical Chymist*.[30]

The instrument theory, however, predated both Bohn's *Dissertationes* and Boyle's experimental critique of chemical principles. Academic chemist/physicians in Germany had developed this approach to chemistry during the middle decades of the seventeenth century. The first mention of the chemical "instruments" can be found in Daniel Sennert's *De Chymicorum cum Aristotelicis et Galenicis Consensu ac Dissensu Liber* (1629 edition). In this work, Sennert described fire, air, water, and menstrua (but not earth) as natural media through which chemical operations were effected.[31] This notion of instruments arose within the context of bringing traditional Aristotelian philosophy into line with chemical practices of analysis and synthesis. Sennert began his academic career at the University of Wittenberg in the 1590s as an Aristotelian, but by 1619 he had amended his Aristotelian theory of matter with notions from new philosophical interests: atomism and alchemy. The key to this change was his investigation of a medieval scholastic argument stating that the Aristotelian elements (fire, air, water, and earth) that entered into a "mixt" through synthesis were not the same ones produced from that same mixt by analysis. Their forms had to be destroyed and recomposed, because the properties of the mixt were not identical to those of the original elements. Yet chemists had shown that some substances could be changed drastically (i.e., made into new mixts) but later recovered in their original form. One pertinent example of this phenomenon was the behavior of metals subjected to dissolution by mineral acids (i.e., menstrua). Certain metals could be converted into salts by the action of acids, but under proper circumstances they could be recovered in their original metallic forms. As William Newman has shown, this fact suggested to Sennert that only the union and dissolution of property-bearing particles, not Aristotelian elements whose properties were defined by forms, could account for this phenomenon.[32] The chemical instruments, only briefly noted in Sennert's work, seemed to be products of his new, hybrid theory of matter. They functioned as the efficient cause of chemical operations, which effected the observational changes in matter through the concretion and dissolution of atoms. In effect, the instruments grew out of a context in which Sennert was increasingly skeptical about the ability of Aristotelian elements to explain chemical change.

The first chemist to elaborate on the chemical instruments in detail was Sennert's onetime student Werner Rolfinck (1599–1673). As a professor on the medical faculty at Jena, Rolfinck first gave chemistry courses in 1637 as the "director of chymical exercises" (*Exercitii Chymici*).[33] He discussed the chemical instruments systematically in his *Chemia in Artis Formam Redacta*

(1661). As suggested by its title, the goal of this work, following Libavius, was to transform chemistry into an academic art. Much of Rolfinck's text presented formal academic arguments in the form of dissertations *about* chemistry. In these dissertations, he strove to define and analyze the nature, principles, and limits of the chemical art in order to place chemistry upon a traditional, academic foundation. Thus, the first book of the text aimed to define the art and examine its aims and "objects." When he came to the objects of chemistry, meaning the substances upon which the chemist worked his operations, Rolfinck entered into an extended academic discussion of the chemical principles. In several sets of academic theses, he extended the arguments of his mentor, Sennert, to include a critique of the various systems of principles: the mercury/sulfur dyad of Geber, the Paracelsian *tria prima*, and the five-principle system (salt, sulfur, mercury, phlegm, and earth) derived from the *tria prima*. He presented arguments in support, and then critical, of each system. Rolfinck ultimately endorsed Sennert's position, noting the difficulty in deducing certain qualities of bodies, such as odor, taste, color, and inflammability, from the supposed composition of their principles or elements. He also suggested, citing an earlier argument from Joan Baptista van Helmont, that chemical principles as substances did not exist in mixts but were the products of chemical operations.[34]

Once Rolfinck had disposed of the principles, in book 2 he examined chemical operations in terms of their causes and effects. Their causes were, of course, the chemical instruments, including the four instruments of Sennert—fire, water, air, and menstrua—and a fifth, earth. Each of the instruments had a well-defined function in chemical operations, but Rolfinck still used Aristotelian language and concepts to describe their properties. For example, the five natural instruments were "active," whereas chemical apparatus were "passive" instruments. Fire produced all the effects of heat, including the expansion and contraction of objects as well as their "generation and corruption." Air served two functions. In its "wet" mode it communicated liquids from place to place, but in its "cold" mode it constrained and congealed salts and oils. Earth was used in certain operations to resolve products. Finally, water and menstrua dissolved or extracted soluble substances from mixts.[35]

Versions of the instrument theory can be found in the works of several prominent physicians and chemists in the last third of the seventeenth century. Many of these chemists had university training and had spent some time at the University of Jena. In addition to Bohn, who studied at Jena during the 1660s, Georg Ernst Stahl (1659–1734) discussed the chemical instruments in his early lectures on chemistry. Stahl received his medical degree from Jena studying under Georg Wolfgang Wedel (1645–1721), who was in turn a student of Rolfinck's. Stahl's textbook in chemistry, *Fundamenta Chymiae Dogmaticae & Experimentalis* (1723), was actually a compilation of student notes from his lectures in Jena during the 1680s.[36] Following the spirit of Rolfinck, Stahl's textbook was a methodized account of the chemical art, which rejected the traditional chemical principles and attempted to build a new theoretical foundation for chemistry based on the motion and accretion of particles and the instrument theory. Stahl's "instruments of operation" were identical to Rolfinck's (fire, water, air, earth, and menstrua), and he provided illustrative, practical examples to explain the effects of the instruments on various species of matter.[37] What emerged with the instrument theory was an alternative approach to chemistry, which had its home within an academic context. All of the chemists who promoted the instrument approach expressed skepticism toward the traditional notion of chemical principles, and they offered the instrument theory as a novel way of conceiving chemical action. In effect, the instrument theory suggested an alternative theoretical framework for chemistry.

Boerhaave extended the instrument theory far beyond previous versions by using the instruments for the taxonomic and pedagogical purpose of ordering various empirical facts and theoretical concepts relating to chemistry. In his course, the instruments acted as pedagogical loci around which relevant chemical phenomena were organized and discussed. Under the topical heading of each instrument, Boerhaave presented a type of "natural history" in which he defined the characteristics of the instrument and described the various forms and actions of the instrument though paradigmatic examples. For instance, fire was the first and primary instrument in chemistry. He defined "fire" as "the everlasting, the most penetrating, and the most pure" substance, which "remain[ed] in the swiftest motion through the attrition of its solid parts."[38] He explained that there were two types of fire: "shining" (*lucens*), which he associated with light, and "burning" (*urens*), which involved combustion. He maintained that the burning type of fire was produced by the motion of fire particles in bodies, but he did not present a detailed, corpuscular account of this process. Instead, he listed scenarios in which motion induced the action of fire. In one example he described the sparks and heat that were generated through the rubbing of a flint on steel or by the concussion of a hammer on iron. Another specifically chemical example was the heat and flame produced by the violent motion of oil of vitriol in dissolving a volatile (vegetable) oil.[39] In each example, the point was to show that the subtle instrument, "fire," acted as the cause of the effects generated, whether they were heat, motion (i.e., ebullition), flame, or light. Boerhaave followed the same structure in describing each instrument: definition followed by a discussion of the forms that the instrument could take and effects that each form caused.

Boerhaave's pedagogical method was to order his examples from simple to complex in order to build chains of inference, which supported his theoretical claims about the more complex phenomena. The examples just mentioned as types of "burning" fire were designed to demonstrate that "fire" was, in fact, a subtle, weightless fluid that interacted with normal matter to produce all the empirical effects associated with common fire. As he had for much of his first course, Boerhaave borrowed several of his examples for "burning" fire from Johannes Bohn's *Dissertationes*.[40] Yet he derived his concept of fire from a local source: the physics course of his former mentor Wolfred Senguerd (1646–1724). In his textbook *Philosophia Naturalis* (1680), Senguerd described "fire" as a ubiquitous fluid composed of "subtle particles" present in all space, even vacua, and capable of insinuating itself into the pores present in normal matter. The motion of this fiery fluid was induced through "attrition" (i.e., the rubbing or collision of bodies) with normal matter or other fire particles and caused the phenomena of heat, fire, and light.[41] Boerhaave structured his examples of "burning" fire to guide his auditors to the conclusion that each form of burning fire depended on the induced motion of fire particles. In the most basic example, the rubbing of flint on steel, the source of the motion and its effects, heat and the generation of sparks, were easy to understand in a straightforward and observable manner. Each subsequent example was slightly more complex, but the pedagogical point was that each situation was a version of the original example. In the final, properly chemical example, the mixing of acid and oil, the heat generated by the violent interaction of two chemical species had the same cause as the other examples: the agitation of fire particles by the motion of normal matter. Thus, in addition to providing the student with a taxonomy of the effects of fire, Boerhaave also provided a manner of arguing with empirical examples for the existence of his subtle fire and its effects in chemical interactions.

Boerhaave included some topics in his course solely for their theoretical interest. Most of these topics examined novel phenomena, new methods of chemical analysis, or operations, which shed light on philosophical questions regarding the properties or composition of matter. One was unlikely to find a discussion of these topics in the didactic chemistry textbooks of the day because these operations and phenomena did not produce useful chemical products. Boerhaave's examination of "luminous" fire exemplified this type of topic. The relationship between fire and light was the subject of ongoing debate among natural philosophers during the seventeenth century. Boerhaave remained agnostic about the precise relationship between fire and light in his early chemistry lectures, even though his teacher, Senguerd, argued for their equivalence.[42] Nevertheless, he included a section in his course in which he discussed operations performed with "catoptric and dioptric burning instruments." Boerhaave's discussion of this topic stemmed from his study of the work of Samuel Duclos (d.1685) and Wilhelm Homberg (1652–1715) at the Académie royale des sciences. In 1667 Duclos used a burning mirror to calcine samples of antimony metal (i.e., regulus of antimony), lead, tin, and sulfur, after which he noticed that each exhibited an increase in their weight. Later, in 1702 (the same year as Boerhaave's first chemistry lectures), Homberg performed several experiments on samples of gold and silver using the Duke of Orleans' large burning lens, designed by Ehrenfried von Tschirnhaus. Boerhaave was enthralled by the power of these optical instruments to concentrate fire "quickly and most vehemently" on bodies placed in their foci. These instruments not only revealed the properties of fire, but they also seemed to offer a novel method of chemical analysis. Homberg, in fact, had initiated a research program centered on the burning lens through which he hoped to reveal the elemental composition of metals and minerals.[43] Boerhaave's inclusion of a discussion of these instruments revealed his philosophical interests and the theoretical focus of his lecture course. Operations performed with these lenses did not generate useful chemical products but rather generated knowledge about the nature of fire, light, and (ideally) the composition of bodies.

Boerhaave's examination of fire also illustrated the limitations of the instrument theory. The action of the instruments alone could not account for all common chemical phenomena, and so he reintroduced into his system of chemistry entities that performed some of the same functions that the chemical principles had previously. One such phenomenon that could not be explained by the action of fire alone was combustion. The problem within Boerhaave's chemistry was that not all substances began to burn when exposed to sparking, flame, or intense heat. Combustibility seemed to be a property determined as much by the composition of the object as by the proper application of fire. In the didactic tradition, this phenomenon was explained by the chemical principle "sulfur." Every combustible body had a certain degree of sulfur in its composition, which allowed the body to generate flame and combust under the proper circumstances. In effect, the sulfur principle functioned as the material cause of flame.

By the beginning of the eighteenth century, chemists had devised numerous theories of inflammability and combustion that incorporated some version of the sulfur principle. Many of these theories had extended the sulfur principle far beyond its origins in the mercury/sulfur dyad or in the *tria prima*. Homberg, for example, argued that sulfur was the primary, active principle that caused both the corrosive properties of acids and the combustion of inflammable substances. In Homberg's system, sulfur in its purest form was the "matter of light," an assertion he used to explain the light emitted by fire as well as the perceived weight gain in metals when calcined in the focus of a burning lens. In the latter case, Homberg argued that particles of sulfur as light broke apart the particles of the metal and combined with them, increasing the weight of the whole.[44] Georg Ernst Stahl also derived his famous phlogiston theory from the traditional notion of the sulfur principle. He asserted in his *Zymotechnia Fundamentalis* (1697), a book detailing a novel theory of wine and beer making, that fermentation was a form of combustion. The sulfur or phlogiston—Stahl used the two terms interchangeably—released during the fermentation of vegetable matter gave the products of fermentation (wine and beer) their pleasant aroma and prevented them from going stale. In an appendix to the main text, he argued that the characteristics of "sulfurous" bodies such as oils, grease, and mineral sulfur itself derived from the presence of phlogiston. Over the next two decades, Stahl applied his phlogiston principle to explain a wide range of chemical phenomena: inflammability, the transformation of metals to calces, the color of bodies, and the action of acids.[45]

Following the lead of his contemporaries, Boerhaave accepted a version of the sulfur principle into his chemistry to explain the selective combustibility of substances. Following Bohn's solution to the problem, he posited an entity to act as the cause of combustion. He called this entity the *pabulum ignis*, or food of fire. As he explained, through its consumption fire "is conserved, spread, and increased."[46] Boerhaave did not describe the mechanism by which the *pabulum ignis* maintained combustion but only provided an empirical example of its existence. He described how "the purest linen" burned quickly and how "pure mineral sulfur"—a form of the *pabulum*—may be extracted from its ashes. Materials rich in the *pabulum* were also substances traditionally considered to be "sulfurous," such as vegetable oils, charcoal (*carbo ligneus*), bitumen, and coal (*carbo fossilis*).[47] Unlike his contemporaries Homberg and Stahl, Boerhaave limited the role of his *pabulum* to the material cause of inflammability. As such, *pabulum ignis* had no role in the fermentation of sugar, calcining metals, or the action of acids. For Boerhaave, each of these phenomena constituted a different type of chemical process. Each involved the generation of heat, which was the result of normal matter in motion inducing motion in particles of fire. Only the *pabulum*, however, induced violent motion in particles of fire merely by its presence, which in turn generated heat, light, and flame during combustion.[48]

Boerhaave approached the other four instruments in the same manner that he examined fire—by elaborating their defining characteristics and usefulness for chemistry. He defined air as consisting of "the most minute, solid, and mobile bodies" that were "fluid, weighty, and elastic."[49] The two main properties of the air relating to its role as an instrument of chemistry were its elasticity and its ability to act as a medium for carrying "exhalations": animal, vegetable, and mineral vapors that it took up from a variety of places. The air's elasticity, or ability to exert pressure, was a primary locus of experimentation and topic of demonstration lectures for the Leiden professors of physics and derived directly from Robert Boyle's experimental work with his air pump on the subject.[50] In his early lectures, Boerhaave argued that the elasticity of air played a necessary mechanical role in fermentation and combustion by "holding together, moving, and rubbing against" the particles of fire during these operations.[51] The air's elasticity and its ability to act as a transporting medium often acted in tandem. For instance, Boerhaave asserted that the air both circulated subterranean exhalations necessary for mineral formation and, through its elasticity, forced the exhalations into proper subterranean pores, where they were concreted into minerals.[52]

Water and earth were instruments in Boerhaave's scheme because they also had the ability to act as media for other substances. He defined water as a fluid substance whose particles were in constant motion but was not elastic like air. Water acted as a medium for volatile salts, such as acids and alkalis, which it dissolved as a type of menstruum or solvent. Earth was a solid substance that possessed no innate motion. Earth acted as a medium by binding or "fixing" saline and oily substances. A salt fixed in earth would not dissolve in water and during distillation would remain as an insoluble mass at the bottom of the retort. A salt or oil that was not fixed in earth would be volatile and during distillation would either fly off into the air or be dissolved in water.[53]

The last natural instrument in Boerhaave's chemical theory was the menstruum. The term menstruum traditionally referred to acidic solvents and was a somewhat archaic term by the turn of the eighteenth century. Boerhaave, again building on ideas from Bohn's *Dissertationes*, expanded the meaning and applications of "menstruum" by focusing on the theoretical mechanisms through which a traditional menstruum acted as a solvent. He defined a menstruum as "that which when applied to bodies divides them into parts, on account of which the whole, whether a fluid or solid, remains unaltered, and then flows or congeals with these [parts]."[54] Boerhaave took this general definition and applied it to a wide range of phenomena in which bodies were seemingly broken apart but could later be recovered by various means. Thus, the category of menstruum included not only liquid solvents but also any type of chemical combination or amalgam in which the initial reactants could be recovered. One paradigmatic example of a solid menstruum was cinnabar, which was a compound of quicksilver and mineral sulfur. Cinnabar could be made simply by mixing and heating quicksilver with mineral sulfur. The resulting compound resisted decomposition in the fire. The original quicksilver, however, could be recovered by various methods, such as by grinding and then heating the cinnabar with iron filings.[55]

Unlike the other instruments, the term menstruum did not refer to a specific chemical species or substance. A menstruum was a specific and latent "power" that a given chemical species possessed, and the power of each menstruum was limited by being able to act on only a specific set of substances. For his course, Boerhaave composed lists of chemical species that could be dissolved by various menstrua. In the section on mineral acids, he explained that oil of vitriol (modern sulfuric acid) could dissolve gold, copper, tin, *regulus* of antimony, zinc, and mercury, whereas aqua fortis (modern nitric acid) could dissolve all metals except gold and tin.[56] In contemplating these facts, he attempted to establish some general principles that accounted for the selective activity of menstrua. He postulated that a menstruum dissociated the body of its "object" by forcing its "active parts" into the pores of the object's body, thereby breaking the body's particles apart. He asserted that the effectiveness of a given menstruum against a specific type of body was determined by four factors: the relative proportion of the menstruum's "active parts" to the body's pores, the exterior figure of the body, the rigidity and solidity of the body, and motion of the menstruum. He provided empirical examples showing the effect of each factor. For example, regarding the proportion of active parts to pores, Boerhaave offered the example of the violent action of oil of vitriol (sulfuric acid) on iron. The more that one diluted the acid, the less violent and the slower the dissolution of the metal was.[57]

Boerhaave was well aware that his corpuscular theory of menstrua was difficult to prove empirically. His explanation of the selective activity of menstrua, that the "active parts" of the menstruum must fit into the pores of the dissolved body, mirrored commonly posited theories found in contemporary didactic textbooks and elsewhere. Both Nicolas Lemery, in his *Cours de chymie*, and Wilhelm Homberg, in several *Mémoires* read to the Académie royale des sciences, used versions of the point/pore model to explain the selective interaction of acids and alkalis.[58] Although this model accounted for the behavior of menstrua (to a limited extent), no theoretical model could predict a specific menstruum's behavior. Because of this, Boerhaave concluded that effects and effectiveness of all menstrua could only be determined by experience. Knowledge about specific menstrua had to be generated through the analysis of chemical operations and systematic experimentation.[59]

Boerhaave attempted to impose order on the various species of chemical menstrua by classifying them based on how they were used in chemical operations. First he distinguished between "dry" and "liquid" menstrua. This division roughly corresponded to the traditional classification of chemical separations between the "dry way"—in which dry reactants were placed in a vessel and heated to fusion or vaporization—and the "wet way"—in which reactants were dissolved in liquid solvents with little or no heat.[60] As examples of dry menstrua, Boerhaave cited amalgams of mercury (i.e., quicksilver) with other metals, cinnabar, and volatile salts fixed in plants, animals, and minerals.[61] All of these examples constituted, for Boerhaave, dry menstrua, because they can be made and also separated without recourse to liquid solvents through sublimation or dry distillation.

The "liquid" menstrua included all of the traditional chemical solvents and any related phenomena in which bodies were broken apart in processes involving water. Boerhaave divided the proper, liquid menstrua into eight categories: aqueous, oily, alkaline spirits, acidic spirits, spirits from vegetable fermentation (i.e., alcohol), fixed alkali, "fossil" salts, and combinations of various menstrua, which produced "various new effects."[62] These eight categories appeared to signify classes of substances that acted as solvents, but again Boerhaave's concept of menstruum extended beyond the notion of a solvent. Under the class of "alkaline spirits," he explained how alkalis selectively dissolved acids and salts—phenomena that modern chemistry classifies as examples of displacement reactions—along with other types of substances, such as oils, sulfurs, "resins," and "gums."[63] In Boerhaave's theory the action of alkali particles in decomposing these substances unified each of these examples as a single class of chemical menstruum.

OPERATIONS

After Boerhaave had concluded his discussion of the chemical instruments, he proceeded to the final section of his lecture course: chemical operations. In the didactic tradition, the persecution of chemical operations in the context of recipes was the major part of any chemistry course or textbook. In the public courses, chemists often demonstrated key operations in front of their audience. In Boerhaave's operations course, which he first offered in winter 1703, he presented a version of this traditional didactic course, describing and demonstrating chemical operations. Yet in his chemistry lecture course, he discussed chemical operations throughout the second half of the course but did not demonstrate any recipes. Instead, he chose a few important operations, such as distillation, extraction, and fermentation, which (following his academic method) he defined and divided into subvarieties. All of the operations he discussed in his lecture course were relevant to medical chemistry and pharmacy—particularly the herbal pharmacy of classical Hippocratic and Galenic medicine. Thus, the second half of Boerhaave's lecture course was devoted to plant chemistry and the operations that chemists used to obtain useful pharmaceutical products and reagents from botanicals. These operations were then demonstrated in his "operations" course.

One of the main goals of the "operations" section of Boerhaave's lecture course was to discuss how the theory of chemistry, the knowledge of chemical species and instruments, shaped its practice. This discussion was crucial to understanding the action of operations on plant matter, because as Boerhaave stated at the outset, operations mutated the natural parts of plant matter, thus introducing new forms and scarcely preserving the old.[64] He began his discussion of chemical operations by stating that the operations on vegetables were the "most simple." The most basic operations on plants consisted of those that removed the fluid parts: oils, balsams, waters, and "gums." Of special interest was the removal of the "odorous spirits" or essential oils. Boerhaave explained that these spirits were typically removed though distillation, by mashing the plant and subjecting it to a gentle fire in the retort, which removed the light oils.[65] He cautioned, however, that the action of the fire could "mutate" the parts of the oil, again citing Robert Boyle for a complete discussion of this problem.[66] He suggested, as an alternative to distillation, the use of solvents to extract the oils and other useful substances. The basic menstruum used for these extractions was pure water. Once the oils were extracted with water through "infusion" and "digestion," the oils could then be recovered from the water through distillation with a "gentle" (*leni*) fire.[67]

In the main, however, Boerhaave did not deviate from the typical methods that chemists used to recover chemical products from plant matter. He structured his discussion of operations and their products according to the usual order that a chemist might follow while performing a traditional, destructive distillation. After extracting the "odorous spirits," the chemist recovered various "limpid liquids" from the plant through distillation at greater heat, and finally extracted the alkali salt from the cinders of the combusted plant remains.[68] In keeping with the pharmaceutical aim of his chemistry, he examined both the chemical and medicinal properties of these products, often describing how they could be used as medical remedies. When he came to the topic of "fixed alkali," he described the operational steps for extracting and purifying this salt as a series of theses: how to make the alkali by combusting plant matter, extracting the soluble matter from the cinders with water, purifying alkali through filtration, and storing the alkali in an airtight container. He next described the salt's chemical and medicinal virtues, especially its ability to neutralize acids and to react with oils to form water-soluble sapos.[69] Finally, he included a prescription for how to administer fixed alkali as a remedy: "The best use is when a few grains, v.g. 20 for adults, dissolved in 10 ounces of pure water, is taken on an empty stomach, divided into three equal parts, of which one warm dose may be drunk interspersed between the two other doses, 1/2 hour apart and with mild walking."[70] Boerhaave's prolonged treatment of fixed alkali reflected his ultimate goal: to describe the natural, "chemical" properties of fixed alkali and to show how these properties were deployed both in chemical operations and in medical practice.

Boerhaave devoted extra effort to discussing contentious operations such as fermentation. The chemical understanding of operations that were described as "fermentation" or "putrefaction" was much contested in the seventeenth century. The influential Joan Baptista van Helmont had denoted the term "ferment" to refer to any instance where a vital agent—a *semen*—acted to transmute one chemical species to another. Thus, fermentation could refer equally to the process of transforming sugar into alcohol or transmuting lead into gold. In this context, fermentation was the foundational process of all true chemical change and, as such, was the central theoretical concept in Helmontian chemistry.[71] Boerhaave's views, however, were shaped by a different interpretation of fermentation, which he found in the work of Johannes Bohn and, more recently, Georg Ernst Stahl. This interpretation denied Van Helmont's theory of vital principles and restricted the definition of fermentation to instances in which alcohol or vinegar was produced from plant matter. Most directly influential on Boerhaave was Bohn, who argued that the process of fermentation was the result of the action of fire (the instrument) and water mutating the parts of a chemical species to generate a new substance that was not present in the original.[72] To clarify some of these issues, Boerhaave conducted a series of experiments to test what types of starting materials—honey, wheat, malt, barley, hay—would promote fermentation and what additional materials—brewers' yeast (*flos cerevisiae*) and various salts—helped or hindered the process.[73]

On the basis of his experimental results, Boerhaave presented his views on fermentation according to his standard pedagogical method. First, he defined "fermentation" as a process by which vegetable matter may be changed into alcohol or vinegar.[74] He then offered seven classes of fermentation. Six were arranged according to starting materials—grain, fruit pulp, the "succulent" parts of plants (i.e., leaves, flowers), dried fruit, and honey or sugar—and one was an anomalous example: "waters flowing with ardent spirit weakened after spontaneous fermentation observed by the British." He stated that many "vegetables" had the power to "undertake the motion of fermentation," including juices, brewers' yeast, honey, and sugars. Within the context of his chemistry, the ability to ferment (or cause fermentation) was a latent, chemical property that Boerhaave treated in much the same way that he treated other chemical properties, such as the power of a menstruum to act as a solvent. As such, he did not see yeast as an essential ingredient in fermentation; indeed, there were many examples of spontaneous fermentation and putrefaction of vegetable matter that occurred without the addition of yeast. In chemical operations (and brewing), yeast acted as an agent to promote fermentation into a specific product: spirit of wine. Boerhaave concluded his discussion of fermentation in chemistry by stating the conditions that promoted or hindered fermentation and described the products of fermentation in detail.[75]

Boerhaave's operations course, in which he demonstrated the chemist's techniques, was the part of his chemistry program that was most like a traditional, didactic chemistry course. He taught his first operations course in the early months of 1703, just after he had completed his lecture course for the second time. The site for the course was probably the small laboratory that he maintained next to his house in Leiden. There he presented 108 separate operations on a wide variety of chemical species, many of which he demonstrated to his auditors. In this course, as in many good didactic courses, Boerhaave divided up the operations according to the type of chemical species on which each operation was performed. His basic categories were the three kingdoms of natural history. Operations on plant matter—distillations, extractions, and so forth—came first, followed by operations on animal products, such as the making of soaps (from animal fats) and operations on the salt and "spirit" of urine (i.e., modern urea and uric acid). Finally, he moved on to the mineral kingdom, presenting operations to produce and utilize alkali and acid salts, (mineral) sulfur, antimony (stibnite), and then the metals.[76]

Ursula Klein has argued that Boerhaave's organization and approach to operations followed Francis Bacon's notion of an "experimental history": an inventory of extant operations in the arts and crafts which "complimented natural history." Within Bacon's program of obtaining "philosophical" knowledge from the work of artisans, Boerhaave's operations exhibited the powers of nature as revealed through the manipulation of chemical species by art. Klein persuasively shows that the demand in the Dutch Republic for practical knowledge was immense and that Boerhaave's operations course and the "operations" section of his *Elementa Chemiae* (1732) was directed, in part, toward serving that demand.[77]

While Klein makes a good point about how some may have read Boerhaave's work, the specific order of his operations revealed their primary academic and pedagogical purpose. The aim of the operations course was not to show recipes but rather to display to novices the basic techniques of chemistry and the principles that guided them. In the course, the individual operations were important primarily because they demonstrated general techniques and principles rather than specific chemical products. Years later, when Boerhaave composed the "operations" section of the *Elementa*, he reflected on how his pedagogical approach deviated from that of the typical textbook. He complained that earlier texts often "described [chemical operations] without any manner of order," and thus operations were presented in an irregular manner that gave no hint regarding the design of the art. He stated that when he first composed his course, he resolved to give his pupils "examples of all the chemical operations in such a manner that . . . nothing should be left out that was of Consequence to be known, nor anything be added that was not necessary."[78] To accomplish this, he established some guidelines for presenting his operations. First, no type of operation would be repeated if it could be understood with one example. Second, it would be necessary to demonstrate with "ocular proof" any operation that had something peculiar to it. Third, the operations would be ordered according to the "method used by the Mathematicians," in which the first operations exhibited were those that were necessary to understand or perform those that followed. This method, argued Boerhaave, was the most efficient way to convey the principles and uses of the art. His choice to begin with plant chemistry reflected this pedagogical aim. On the basis of the principle that one should begin with the most simple and proceed to the more complex, he argued that operations on "vegetables" were the simplest to perform and easiest to understand in terms of their principles of action. Operations on "fossils" required the most skill to perform and were the most difficult to understand, so they should be addressed at the end of the course.[79]

The first "operations" that Boerhaave demonstrated, those that he felt were the simplest and most basic, were the three primary methods of collecting plant products from crude plant matter. First, Boerhaave set up his still and placed wormwood leaves (*absinthium*) and water in the alembic, heated them gently, and pointed out the vapors that were generated and condensed in the receiver. These condensed vapors were the "odorous spirits" or essential oil of the wormwood. Then he performed a "decoction" or solvent extraction on the residue left behind from the previous distillation. He added water and heated this mixture to just under its boiling point, pouring off the water and heavier oils extracted by this process. This operation produced an "infusion" of absinth. Finally, the remaining matter in the still was dried and combusted into cinders. He then demonstrated the proper manner to extract the alkali salt from these cinders with water and purify them.[80] In effect, Boerhaave performed for his auditors the processes that he had earlier described in his lecture course. As was the case with that course, he did not present this demonstration as a recipe for its own sake but rather for teaching the techniques and proper theoretical understanding of the operation. That he was operating on wormwood was not as significant as the fact that he was outlining the general technique for extracting oils and demonstrating the power and limitations of "fire" and aqueous solvents in performing this operation. In fact, a few years later he switched from wormwood to rosemary for this series of operations.[81]

As the operations on wormwood suggest, the secondary aim of the course was to reinforce the theoretical explanations covered in Boerhaave's chemistry lectures. This aim reflected the traditional academic division between theory and practice as seen in the medical curriculum at Leiden. The operations course presented the *practica* of chemistry, while the lecture course covered *theoria*. In the *Elementa*, Boerhaave assumed that the reader of the operations section of the text (in volume 2) was already familiar with the material covered in the theory section (in volume 1). For example, in the introduction leading up to his discussion of plant chemistry, he referred the reader back to the earlier, theoretical material, stating, "we shall suppose you are acquainted with all of those things which were explained to you in the *Theory of the Art*."[82] One of the goals of the operations section was to demonstrate through example how the chemical instruments affected matter. Observing the operations performed for them would allow students to recognize effects of instruments and also to better comprehend the properties of chemical species. Boerhaave described several processes designed to combust plant matter and extract alkali salts from the combusted ashes. In describing the products of these processes, he initiated an extended discussion of the properties of these salts. For example, he described their properties as menstrua—their ability to act as a "magnet to water" as well as their attraction for acids, which displaces metals in acidic solutions.[83] Because alkali salts obtained through various methods and from various types of plants all share these properties, Boerhaave pointed out that these salts must be products of the action of fire on plant matter, rather than present in the plants themselves before combustion.[84] This assertion supported one of the main theoretical contentions from the theory course—Boyle's critique of analysis by fire.

The operation course, taught in tandem following his lecture course, established a regular chemistry curriculum at Leiden, which followed the pattern of the medical and the physics curricula. He taught both courses each year from 1705 to 1711, and from then on about every other year until 1728. As he grew more experienced with teaching and doing chemistry, he expanded both courses. When he taught the operations course for the second time in 1705, he included 151 operations, and by 1710 he had raised that number to over 200.[85] In the lecture course, he periodically updated and added to his presentations. For example, when he began to use Fahrenheit's thermometers in his chemical work after 1718, he reworked his lectures on fire to reflect this.[86]

CONCLUSION

In March 1702, as Boerhaave was in the middle of this first chemical lecture course, the Dutch stadholder, William of Orange, died in England. This event ultimately led to the appointment, on May 6, of Jacob le Mort (1650–1718) as professor of chemistry on the Leiden medical faculty. In most years, at least early on, Le Mort presented his course in chemistry, which Leiden students could attend without paying lecture fees. (These were covered by their matriculation fees and filtered to Le Mort though his salary.) Boerhaave, however, refused to give up his chemistry courses. In the winter following Le Mort's appointment, Boerhaave taught his lecture course again and added his first demonstration course of chemical operations. He taught his two chemistry courses every year until 1711, at which point he taught them every other year until 1718.[87] During this time, Boerhaave and Le Mort competed for students and the prestige that came with being a successful and revered professor. Despite the fact that students attending Boerhaave's courses had to pay him lecture fees, he won this competition. The majority of students preferred Boerhaave's courses to Le Mort's.

One reason for their preference could have been the personalities of the two professors. Most anecdotal accounts portrayed Boerhaave as personable and steady, whereas Le Mort, at least after he became professor of chemistry, was usually seen as irascible, unreliable, and somewhat paranoid.[88] A more concrete answer was that Boerhaave was much more successful than Le Mort in integrating chemistry into the medical curriculum. Boerhaave strove to shape his chemistry courses to fit the pedagogical and philosophical norms of the Leiden medical faculty. He based his courses on models of chemical philosophy and practice, Boyle and Bohn, which could be easily integrated into the academic practices of the medical faculty. Le Mort also attempted to integrate his chemistry into the medical curriculum. In his first several years as chair of chemistry, Le Mort acted as *promotor* for at least two inaugural disputations (i.e., dissertations written for an M.D.) on chemical topics.[89] Nevertheless, his chemistry courses were modeled on the traditional didactic chemistry course, which focused on conveying recipes for medicaments and other commercial products. Some of Le Mort's books included extended theoretical discussions of chemistry, but Le Mort's chemical theory relied on a rigid Cartesian matter theory, which had been out of favor in the Leiden medical faculty since the 1680s.[90] Little of his work addressed the usefulness of chemistry for medicine outside of pharmacy. As we shall see in the next chapter, Boerhaave integrated chemistry into his medical courses by deploying chemical concepts and techniques in his discussions of human physiology and pathology. Ultimately, Boerhaave was successful because he was first a physician who developed a clear vision of how the chemical properties of bodily substances shaped human health and disease.

Boerhaave initiated a revolution in the pedagogy of chemistry. In his early lecture courses he rejected the didactic tradition and devised a new method of ordering chemical entities and phenomena. Like the medical curriculum, Boerhaave's chemistry curriculum examined the theory of chemistry first. This theory was shaped by the instrument theory, which defined the natural tools of the chemist arranged according to strictly defined taxonomic categories and described the chemical phenomena derived from these tools in terms of theoretical mechanisms and precepts. Then he moved on to operations, which were ideally designed to teach the student how to apply the precepts to the practice of chemistry. Through this new method, the precepts and definitions of chemistry could be deployed in medical courses, as discussed in the next chapter. Boerhaave created a chemistry, tailored for the medical faculty, that was about *knowing* things—the chemical properties of matter, the kinds of entities that existed in the world, the mechanisms of natural change—in addition to making them.

CHAPTER FOUR

Chemistry in the Medical Faculty

In September 1718, the professor of botany and medicine at the University of Leiden, Herman Boerhaave, presented an oration to the university community on the occasion of his inauguration to the chair of chemistry, his second on the medical faculty. In this oration titled *De Chemia suos Errores Expurgante*, he lambasted some of the systems of chemical medicine that had been popular in the preceding century, notably that of the illustrious Leiden professor François de le Boë, more widely know by his Latin name, Franciscus Sylvius. Boerhaave asserted that Sylvius's medical system was based on the overly simplistic assumption that simple chemical reactions performed outside the body modeled the body's complex physiological mechanisms. For example, he pointed out that because distilled acids mixed with essential oils in the laboratory generated a "fervid" heat, the chemical physicians argued that the acid found in chyle mixed with the "balsam of blood" (an oil generated from blood by chemical analysis) accounted for the warmth of the body. Similarly, stronger mixtures of these two substances caused "burning fevers."[1] To emphasize the naiveté of this position, Boerhaave poked fun at some of the anthropomorphic language deployed by chemical physicians, who described acids and alkalis as "pugilists" who battle and then seek shelter from their enemies in various parts of the body. Following this account, he laconically asserted: "Do you feel I am fashioning a fable and narrating a sick man's dreams and delirious fancies? Yet in recent times physicians have indeed seriously argued that the natural processes of life come about in this way—attacking with witty ridicule the opinions of the Ancients, who actually showed more discernment than their critics." Boerhaave commented that one may learn the basic precepts of Sylvius's chemical medicine in just one hour. One needed only to learn a little about the natures of "acid" and "alkali" and to realize that the whole endeavor involved achieving an "equilibrium of forces" between the two.[2]

Despite this criticism of Sylvian chemical medicine, Boerhaave himself was a great advocate of chemistry in the medical faculty, which was why he was named chair of chemistry in 1718. A perusal of his medical writings and courses indicates the importance of chemical analysis in his medical system and his allegiance with some aspects of chemical medicine. One example is Boerhaave's description of "cystic bile," found in his textbook of medical theory, *Institutiones Medicae*. Within the context of his discussion of digestion, he described this bile as correcting "acidities" in the digestive tract but observed that it was itself neither alkaline nor acid. It was composed of saline, oily, and spirituous parts, but the nature of its action was saponaceous (i.e., soaplike), "disposing oil to mix with water" and dissolving "resinous, gummy and tenacious substances, reducing them to a uniform mixture."[3] Clearly, Boerhaave was not opposed to using simple chemical concepts such as acid, alkali, or "saponaceous" to describe the properties of bodily fluids and their role in physiological processes. In this example, he stated that the cystic bile was neither an acid nor alkali, an assertion which suggests that he did not reject Sylvius's categories of analysis. In fact, Boerhaave embraced the concepts of "acid" and "alkali" as terms useful in defining specific chemical properties and types of substances bearing those properties. In addition, by describing the cystic bile as "saponaceous," he was in fact employing another chemical property that alluded to the bile's function in digestion—as a chemical menstrum that dissolved oily and resinous substances and allowed them to mix with aqueous fluids in the digestive tract.

Boerhaave advocated for a reformed chemical medicine based on, as he saw it, a more limited and empirically grounded notion of the chemistry of bodily substances. This reformed chemical medicine rejected many of the claims of older iatrochemical systems as it embraced others. The roots of Boerhaave's chemical medicine can be found by delving into the extant medical traditions and practices of the Leiden medical faculty while he was a medical student. These medical practices promoted empirical and experimental investigation, while rejecting what Boerhaave saw as dangerously speculative approaches, such as Sylvius's chemical medicine. Yet despite his harsh rhetoric against chemical medicine, he engaged with Sylvius's system, critiqued what he took to be its shortcomings, and adopted useful aspects of it in his own medical practices. Boerhaave tailored the claims of chemical medicine to fit his medical philosophy and pedagogical needs, and he expanded upon it to create an academic space for chemistry in medicine and medical education. For Boerhaave, chemistry was a necessary tool of analysis and theorizing in physiology, pathology, and therapeutics.

Ultimately, this chapter depicts Boerhaave as a chemical physician. This interpretation revises the traditional view of Boerhaave's medicine, which historians have nearly always described as "mechanical."[4] The traditional view is certainly a reasonable one in that Boerhaave himself often stated that his medicine followed the "mechanical method."[5] Recently, however, Rina Knoeff has challenged this view, arguing that Boerhaave saw chemistry, not "mechanics" or "physics," as the foundational discipline for interpreting human physiology. Further, she asserted that during the course of his academic career he shifted from a "mechanical" approach to medicine to one that was decidedly "chemical."[6] I generally side with Knoeff on the chemical character of Boerhaave's medicine, but I have reservations regarding the distinctions that she and other historians have made between "chemical" (using chemical properties to explain phenomena) and "mechanical" (depicting the body as a machine) approaches. As Anita Guerrini has shown, there were many varieties of "mechanical" medicine, some of which incorporated "chemical" principles and properties into their conception of the body.[7] I prefer to examine how Boerhaave defined "chemical" and "mechanical" or "physical" approaches in practice, that is, how he made distinctions within his medical courses and curriculum, and how he deployed chemical knowledge in practice. In the end, Boerhaave devised a new form of chemical medicine, which he sought to distance from the perceived excesses of Sylvius and other iatrochemists, yet he also sought to integrate chemical knowledge into an established, "mechanical" view of the body and its workings. The effect of this integration was that chemistry became institutionalized within his medical curriculum, and by doing so, the status of chemistry as a medical field was raised accordingly. The medical student simply could not hope to master Boerhaave's medical system without a basic knowledge of chemistry and chemical analysis.

PURGING CHEMISTRY'S ERRORS

More than any Leiden professor before him, Boerhaave made a concerted effort to integrate chemistry into the medical curriculum. As early as his first academic oration at Leiden in 1701, he asserted that he "loved" chemistry and suggested that the art was both suitable and useful for medicine.[8] Just how chemistry could fit into medicine and the medical faculty, however, was a complex question that, for Boerhaave, was shaped by extant theoretical, practical, and pedagogical traditions in Leiden. Part of his effort involved justifying the inclusion of chemistry in the medical curriculum and showing that his new, academic chemistry provided knowledge that was useful and necessary for medicine. Chemistry as pharmacy, of course, had always been useful to medicine, but as I argued in chapter 2, knowledge of the chemical properties of medicaments and the operations that apothecaries employed to make them was not a central part of the Leiden medical curriculum. Boerhaave, by contrast, argued that chemical knowledge informed the core subjects of the medical curriculum: physiology and pathology. Thus, chemistry was integral to the physician's understanding of the human body and therefore shaped the proper practice of medicine in a fundamental way.

In the early eighteenth century, Boerhaave's contention that chemistry was integral to medicine, especially for understanding physiology and pathology, was not universally accepted among university physicians. Even in Leiden, one of the few universities to have a chair of chemistry on the medical faculty, chemistry had been cast as an ancillary practice to medicine, and the professor of chemistry generally had the lowest status among the medical professors. Contemporary university medical faculties generally supported this assessment of the status of chemistry.[9] Medical practitioners who shared Boerhaave's opinion that chemistry shaped medical practice flourished largely outside of the university. One reason for this situation was that the founder of the "chemical" approach to medicine, Paracelsus (d. 1541), and his chief revisionist in the seventeenth century, Joan Baptista van Helmont (d. 1644), presented their chemical medicine as a superior alternative to academic, Galenic medicine, which they portrayed as wrongheaded.[10] For much of the seventeenth century, the resentment went both ways. In Paris, for example, the Galenic, Parisian medical faculty actively blocked the appointment of Paracelsians to the faculty, forcing the chemical physicians to use the royal court and court-sponsored institutions as their power bases.[11] Chemistry had more success in German schools, notably at the University of Jena.[12] These successes, however, were often the result of noble patronage rather than acceptance by the university community. Johannes Hartmann (1568–1631), for example, was named professor of *chymiatria* at the University of Marburg in 1609 at the behest of Landgrave Moritz of Hesse, the university's patron. Hartmann taught his students and the landgrave's court Paracelsian remedies and theory, but he had to endure bitter attacks from both his colleagues and critics outside the university.[13]

In Leiden, Franciscus Sylvius (1614–1672) attempted to forge a moderate chemical medicine that synthesized chemical theories, purged of their Paracelsian excesses, and Harveian anatomy with the Galenic medicine that was firmly entrenched at the university.[14] Sylvius's discussion of digestion in terms of the interaction of acids and alkalis, for example, bore striking similarities to the discussions of digestion found in Van Helmont's *Ortus Medicinae* (1644). Sylvius, however, denied any influence from Van Helmont, arguing that he devised his approach to physiology before the works of Van Helmont were published.[15] For Sylvius, "acid" and "alkali" referred to chemical properties that inhered in the particles of substances and were defined by their mutual affinity for each other.[16] In medicine, these substances included food, medicaments, and bodily fluids. Many of the latter had been only recently identified through anatomical and physiological research. Within this context of rapid discovery, Sylvius built a theory of physiology in which physiological processes, like digestion, were to be understood as the interaction of chemical species. He deployed his chemical physiology within a traditional, Galenic conception of health based on the balance of the bodily fluids or "humors." Some diseases were caused by acidic or alkaline imbalances in the bodily fluids, which could be corrected by reestablishing the balance, that is, treating "acidic" diseases with alkaline remedies and vice versa. For example, Sylvius argued in his *Praxis Medicae Idea Nova* that "dry" diseases, a concept from Galenic medicine, were caused by "acrid biles" that produced alkaline salts in the blood. He suggested that, to remedy this situation, bile salts needed to be diluted and the "urine" (i.e., alkali part) drawn out. Thus, he prescribed remedies made of acid and oil—the acid to counteract the alkali and the oil to dilute the bile.[17]

Chemical physiology, however, was only one aspect of Sylvius's medical program. He was also committed to an empirical and experimental approach to natural philosophy and medicine.[18] He was an active anatomical researcher, a passion that he effectively applied in his medical teaching, since he often allowed his students to participate in his physiological experiments.[19] When he conducted Leiden's *collegium medico-practicum*, he took students to the St. Catherine and Cecilia Hospital to see patients every day, rather than the required two times per week. He also obtained permission to perform autopsies on patients who had died in wards other than the two designated for teaching. This situation greatly increased the number of dissections that his students could observe. A Scottish student, Robert Sibbald, reported that in 1660–61 he witnessed twenty-three dissections under Sylvius's direction, an astounding number for the time.[20] Yet even his chemical physiology had an empirical foundation. In the example of "dry" diseases just presented, Sylvius's recommended treatment was based on chemical interactions that one could observe in the laboratory. Any chemist could show that acids neutralize specific alkalis, and Sylvius's recommendation to dilute the "acrid bile" with oil was based on the well-known chemical practice of "dulcification," in which caustic substances were rendered milder through combination with another, "bland" substance.[21]

Despite Sylvius's immense popularity with medical students at Leiden, his chemical medicine quickly lost favor in the medical faculty after his death in 1672. New medical faculty members rejected or ignored his chemical theories of physiology, while embracing his anatomical and experimental approach. In 1668 Charles Drélincourt (1633–1697) was appointed professor of anatomy, and soon after his arrival in Leiden, he published a short treatise (under the pseudonym Le Vasseur) in which he criticized Sylvius's chemical theory of digestion.[22] Drélincourt continued Leiden's tradition of the experimental study of anatomy and physiology. In addition to conducting human dissections, he offered demonstration lectures on physiology in which he vivisected dogs to exhibit the circulation of the blood and the structure of the circulatory, lymphatic, and digestive systems.[23] In 1687, he was joined by Anton Nuck (1650–1692), who also engaged in experimental physiology, demonstrating to students experiments on the fertilization of the mammalian egg in the uterus of dogs, displaying models of the human lymphatic system, which he created, and performing fine anatomy on glands.[24]

Neither of these professors, however, rejected chemical approaches to medicine outright. In practice, Nuck, Drélincourt, and other Leiden medical professors integrated specific chemical techniques, remedies, and concepts into their medical courses when they proved to be useful. As Harold Cook has pointed out, every capable anatomist and natural historian would have known some basic chemical techniques, which they used to analyze and preserve specimens.[25] In fact, many anatomists outside of Leiden had already begun to incorporate chemical analysis in making physiological arguments. Next to Sylvius, Thomas Willis (1621–1675) in Britain was a pioneer, publishing tracts on digestion, urine, and fever.[26] When Drélincourt taught Leiden's *praxis medica* course, the remedies he prescribed were mainly herbal, but he also incorporated many "chemical" remedies that Sylvius had advocated, such as those made from arsenic, vitriol salts, copper, and antimony.[27] Nuck embraced chemical analysis in his anatomical work, especially when it could help him make arguments in physiology. Nuck's *De Ductu Salivali Novo,* which he published in 1685 while he was a physician and anatomist at the Hague, exemplified this use of chemical analysis. In this work, he conducted fine anatomy and ligation experiments on the vessels and glands near the orbit of the eye in order to argue that the gland under the eye distributed saliva to the mouth, not aqueous humor to the eye. To support his argument, he submitted both aqueous humor and saliva to a battery of chemical tests in order to determine the differing "constitutions" of the fluids. This chemical analysis included distilling the fluids into various components and submitting them to a battery of tests with various acids and alkalis.[28]

The issue for Nuck and Drélincourt was not the use of chemical remedies or techniques in themselves, but rather how the physician applied chemical knowledge. At the core of the Leiden medical faculty's empiricism was a belief that all medical claims must be based on empirically established facts, not on assumptions derived from a philosophical or theoretical system. As Nuck argued in his *De Ductu Salivali Novo*, valid claims in anatomy are based on "one hundred observations and experiments," not philosophical speculation.[29] For Nuck, the aim of medical experimentation was to establish facts about human anatomy and physiology and not to build rigid theoretical systems. This approach was similar to what Andrew Wear has called "the way of the anatomists" to describe William Harvey's approach to anatomical work. Harvey proceeded by demonstrating observable anatomical facts, which he considered to be a more reliable form of knowledge than reliance on extant theories.[30] Although Sylvius undoubtedly saw his work within Harvey's tradition of experimental anatomy, Nuck and Drélincourt deemed that Sylvius had relied too much on theoretical speculation.

Boerhaave's appointment to the chair of chemistry in 1718 signified the successful integration of chemistry into the Leiden medical faculty, since the formerly marginalized subject was now taught by the medical faculty's most prominent member. Nevertheless, he still worked to justify chemistry as a legitimate subject for the medical curriculum. One of his first acts as professor of chemistry was to petition the Leiden curators for permission to give an oration to the university community.[31] That Boerhaave insisted on giving an inaugural oration for his new chair, an honor he could have easily declined, suggests that he wished to exploit the opportunity to advocate for his chemistry program. In *De Chemia suos Errores Expurgante*, Boerhaave argued that previous errors and heresies that many chemists once held had been purged from chemistry by modern chemists. These errors included disreputable practices, such as alchemical charlatanry, but also earlier forms of chemical medicine, such as Sylvius's acid/alkali theory and

Van Helmont's theory of "ferments."[32] In practice, Boerhaave utilized the work of both of these men as resources for constructing his medical system. He intended the oration to be a rhetorical exercise, to distance himself from the perceived excesses of these approaches. The main rhetorical point was that the new chemistry purged of its former errors could now be used to foster progress in medicine by correcting the previous mistakes of chemical physicians.

To set up his argument about how chemistry had corrected its errors, Boerhaave devised an account of how and why previous chemists had originally erred. According to Boerhaave, most of the errors in chemistry (like errors in medicine) came about through the chemists' lack of a disciplined self-restraint in their theorizing. He argued that chemistry originated among miners, who were uneducated and, as a result, guided by superstition. These superstitions were passed on from master to pupil, until during the "hermetic" revival of the Renaissance, "erudite physicians," rejecting Galen and the "Arabs," accepted both the established facts and superstitions of chemistry. In an uncited reference to Paracelsus, Boerhaave stated: "one could even observe most authoritative and eminent chemists in full earnest teaching and inculcating things, which had been created by poets . . . [as if they] actually did exist: Fauns, Satyrs, Genii, Nymphs, Pygmies, demigods, the masters of woods, mountains, rivers, and of subterranean places and air."[33] Within this climate of intellectual speculation, Boerhaave contended that some chemists could not restrain themselves even from interpreting the sacred scriptures as descriptions of "the art of making gold." He quickly distanced himself from such activities, however: "These people defiled one and all with their commentaries, on all occasions prattling about allegories, emblems, images, enigmas—so much so that everything which may be read in Holy Scripture, however, clear, unambiguous, and plain, was distorted by these trifling fools into an inapposite meaning."[34] Boerhaave asserted that such zealotry induced practitioners of the chemical arts to "squander their property," yet these same men often failed to keep within reasonable limits when defending the art against criticism and recounting its capabilities.[35] Thus, the chemists' errors derived from their own superstition and lack of a disciplined methodology, not from any fallacy contained within the chemical art itself.

Even educated, chemical physicians had erred for the same reasons; they constructed universal principles from phenomena that were rightly attributed to individual bodies or valid only in limited situations. This was the root of Franciscus Sylvius's errors in chemical medicine: applying the affinity between acids and alkalis to all chemical action in the body.[36] In another example, Boerhaave pointed to the doctrine of "ferments" as popularized by Van Helmont by presenting a rough account of the origin of this theory. It was well known that the oil of certain "acetous" plants could be converted into a volatile spirit that was soluble in water and flammable (i.e., alcohol). Yet, by a similar method, this oil could be made into an "acetous" spirit that was also soluble in water but extinguished flame (i.e., vinegar). Both of these processes were called "fermentations," but some chemists (i.e., Van Helmont and his followers), based on the fact that a similar process produced these two dissimilar substances, had proclaimed the existence of a universal "ferment," which was responsible for all true chemical changes in every body. Boerhaave maintained that such overgeneralizations were detrimental because they promoted disunity in the chemical community and hindered the progress of the art. Without a rigorous methodology to discipline each practitioner's speculation, individuals tended to construct theoretical frameworks based on their own, limited experience. As Boerhaave lamented, "this is why there are so many sects among those people—starting from different experiments, everyone sets up a general theory which fits with, and is closely determined by, his own observations."[37] The result was dissension and confusion, which served only to tarnish the whole field.

Boerhaave contended that the modern chemists were able to purge chemistry of these errors because they applied the proper method of discovery. He recounted that later chemists learned to restrain their speculation by accepting only those phenomena that they apprehended through their senses. They found that each time they brought new bodies into contact with others, new effects were produced that were not reducible (by themselves) to a generalized law. According to Boerhaave, they realized that "one needs an enormous store of observations, a most cautious scrutiny of this material and, finally, a careful mutual comparison of all data before one is entitled to postulate a universal rule that is valid for all natural reactions."[38] By applying this cautious method to the precepts of chemical medicine, these later chemists revealed the errors of the iatrochemists and established chemistry's proper place among the arts. Boerhaave asserted that these chemists "demonstrated" that the effects of nature were wholly different from the effects of art, and therefore they undermined the foundation myth of chemical medicine by showing that chemical operations did not mirror natural, physiological ones. As he contended, "the human organism deals with the causes of illness through means that cannot be imitated at all by the science of chemistry; . . . life and health depend on so many various, intricate, subtle, and wholly intangible causes." Thus, he concluded, "in the most *prudent* manner chemistry rectified its errors."[39]

The message of *De Chemia suos Errores Expurgante* was that chemistry was useful to medicine because it helped the physician to correct the errors of the chemical physicians. But the usefulness of chemistry had limits. The first limit was the methodology of chemistry. The proper chemical practitioner collected numerous observations and considered all options carefully and critically before positing a general precept. As depicted in *De Chemia*, chemistry followed the same methodological and *moral* rules that Nuck advocated for physiology and that Boerhaave posited for medicine generally (as I outlined in chap. 1). Boerhaave had described this approach to medicine as a "Hippocratic" method of medical practice, a rubric that he now applied to chemistry.[40] *De Chemia* also suggested a second limit, which concerned the kind of knowledge that chemistry provided to the physician and the proper uses for this knowledge within the medical sciences. Boerhaave rejected the iatrochemical notion that the human body acted as a chemical apparatus in which chemical operations in vitro could be simplistically used to model physiological processes in vivo. By the time that he delivered *De Chemia* to the Leiden academic community, he had developed specific ideas about the role of chemistry in the practice and teaching of medicine.

CHEMISTRY IN THE CURRICULUM

One way in which Boerhaave, as an academic, established the relationship of chemistry to medicine was to define the place of chemistry in the medical curriculum, or as it became known, his "method" of medical education. Boerhaave's medical method, in this case, referred to the ordering of medical courses, topically arranged to convey the precepts of the art to a student.[41] In this ideal curriculum, each course built upon the concepts and skills the student had learned in previous courses and added new skills needed for subsequent courses. Thus, the earliest courses established basic skills and the basic theoretical and methodological foundations for Boerhaave's medical system. Later courses built upon these foundations to develop more specific medical skills and concepts and, ultimately, to teach the student how to apply the precepts of the medical system in practice. The skills, concepts, and techniques that chemistry contributed to help the student master medical theory and practice defined its place in the medical curriculum.

According to Boerhaave's method, chemistry, along with physics and mathematics, provided the foundational skills needed for proper medical training and practice. Boerhaave had envisioned his method of medical training early in his academic career He briefly described his ideal medical curriculum in his second academic oration at Leiden, *De Usu Ratiocinii Mechanici in Medicina* (1703).[42] As his prominence and the student interest in his medical courses grew, however, he devised a more rigorous exposition of his curriculum, which he offered as a private lecture course in fall and winter 1710–11. This course, De Methodo Addiscendae Medicinae, presented a detailed discussion of each medical course and the specific topics examined within them. Boerhaave provided an academic definition for each course topic, explained the content of each subject, and *justified* the necessity of each topic for proper training in medicine.[43] Thus, individual topics, like chemistry, were defined by how they contributed to the whole curriculum and how they complimented, yet were distinct from, topics that came before and after them in the curriculum. Chemistry appeared in De Methodo after the study of mathematics, "mechanics," and experimental physics but before the study of botany, pharmacy (as materia medica), and physiology. Chemical knowledge built upon (but was distinct from) the contributions of "mechanics" and physics, and served as a foundation for the study of botany, pharmacy, and physiology.

To understand the role of chemistry in the curriculum, we first need to examine what Boerhaave meant by the terms "mechanics" and "mechanical medicine," which he advertised in his 1703 oration. Boerhaave's understanding of "mechanical medicine" derived from what he took to be the *practice* of the mechanical arts—a notion of "mechanics" that reflects what Jim Bennet has called the "Mechanics' Philosophy."[44] In Boerhaave's mind, the practitioners of the mixed mathematical arts—surveying, engineering, instrument making, and the like—understood the objects of their art through observation, measurement and, ultimately, the mathematical modeling of physical properties. Boerhaave argued that these "mechanics" followed a precise, if often unstated, method of practice, what he called the "method of mechanics" or, on occasion, the "method of the geometers." Following this method, the practitioner first observed his "object" empirically, and then constructed (or, literally, drew) known mathematical entities (i.e., geometrical forms) or measured numerical magnitudes that represented the observed phenomenon. By applying mathematical rules and relations to the mathematical entities, the practitioner drew reliable conclusions regarding the nature and motions of the phenomena he observed.[45] Boerhaave contended that with this approach, the "mechanics" had devised an empirical and experimental method to define, predict, and control the behavior of physical bodies.

Boerhaave believed that the most prominent and reliable natural philosophers, and even a few physicians, had used the method of mechanics to shape their work. In his public pronouncements, he consistently associated important innovations in natural philosophy, such as the work Christian Huygens and Isaac Newton, with this method and credited failures to the lack of its use.[46] The most likely source of Boerhaave's enthusiasm for the method of mechanics was his mentor in the Leiden philosophy faculty, Burchard de Volder. In his lectures on *physica theoretica*, as well as in several academic orations, De Volder argued that mathematics was the best tool to discipline human reasoning in the sciences and, as such, he advocated for a union of the disciplines of mathematics and physics (traditionally separate in the university). In a university oration from 1698, he argued that medicine too should be subjected to a "hypothesis" of mechanical causes, which was to be tested through experiment and the application of mathematics.[47] De Volder was not alone in making the case for mathematical approaches in medicine.[48] In Britain, a group of physicians lead by David Gregory (1659–1708) and Archibald Pitcarne (1652–1713), who identified themselves as followers of Newton, experimented with a similar application of mathematics to human physiology.[49] One of these "Newtonian" physicians, Pitcarne, filled the chair of medical praxis in Leiden for a short time (1692–93) while Boerhaave was a student. During his stay in Leiden, Pitcarne delivered an inaugural oration and conducted several disputations that advocated for what he called "iatromathematics." Many aspects of Pitcarne's approach to medicine, such as his rejection of "ultimate causes" proffered by "philosophical sects" and his application of the "laws of hydraulics" to fluid flow in the body, had striking parallels in Boerhaave's medicine. There is little direct evidence, however, to suggest that Pitcarne had any substantive contact with the young Boerhaave, although Boerhaave was undoubtedly aware of Pitcarne's work.[50]

In De Methodo Boerhaave suggested that the physician begin his medical training by studying the mathematical arts, specifically geometry and "mechanics." He justified the study of the mathematical arts by arguing that because medicine concerned the human body, which was a physical entity, the "science" of physical bodies in general laid the foundation for medicine. The student first examined the characteristics of bodies in general as determined by observation: extension, impenetrability, figure, and motion. Boerhaave argued that because one comprehended bodies by observing and measuring their "figure," the study of geometry and trigonometry was an essential part of the physician's training. Next the student needed to learn how to depict the motion of bodies, which he learned by studying "mechanics." Boerhaave defined "mechanics" as the art through which one "calculated the quantity of motions" caused by gravity, impact, or attraction.[51] The topics under "mechanics" included an examination of basic concepts, such as space, time, and gravity, followed by mathematical formulations of the simple machines (lever, inclined plane, pulley, wedge, and screw) and, later, "hydrostatics": the "gravity" (i.e., specific gravity), pressure, and motion of fluids.[52] The point of this study was to familiarize the physician with the mathematical methods and models that he would use later in the medical curriculum to understand the structure and movements of the human body. In effect, mechanics, especially fluid mechanics, provided the basic interpretive framework for anatomy and physiology.

According to De Methodo, once the student had mastered the mathematical arts, he moved on to "physics," by which Boerhaave meant the study of the properties and motions of bodies through experiment and measurement. This notion of physics was more restricted than the term was commonly understood in the early eighteenth century. Here physics signified the general study of material bodies in nature, a jurisdiction that might include Boerhaave's notion of "mechanics," experimental natural philosophy, chemistry, and parts of medicine.[53] As defined by Boerhaave's De Methodo, physics roughly followed the physics curriculum in the philosophy faculty at Leiden: lectures on the general properties of bodies and their motions, followed by experimental demonstrations designed to elucidate those principles. Boerhaave's curriculum focused on experimental demonstrations, which were modeled on those which Burchard de Volder and Wolfred Senguerd had typically performed in Leiden's *theatrum physicum* while Boerhaave was a student. These demonstrations exhibited the principles of bodies discussed in the lecture course and also showed students how to measure the "physical" properties of solids and fluids—weight, volume, density, viscosity, and so forth—in order to understand how these properties shaped motion.[54] Boerhaave used these physics demonstrations to demarcate the pedagogical role of "physics" from that of "mechanics." Whereas mechanics taught the skill of mathematical modeling, physics taught the skills of *measuring* the observable properties of bodies, which included the operation of measuring instruments: barometers, thermometers, balances, and the like. As such, physics was an experimental and empirical enterprise. As Boerhaave stated, physics "describe[d] all actions, which are exerted between bodies, which may be noted accurately by the power of the senses." Thus, physics concerned the empirical observation of properties of bodies and those non-observable properties of bodies that could be deduced "by reasoning from observation."[55]

The function of chemistry in De Methodo was to investigate the specific properties of bodies, which the techniques of neither physics nor mechanics could reveal. According to Boerhaave, these properties were of a "singular nature" and "depend upon particular actions that are able to be known only through [chemical] experiments." As examples of these properties, he identified the selective "attractions" and "effervescences" that chemists observed when they brought specific chemical species into contact with other chemical species. Boerhaave understood these effects as a kind of motion, specifically the separation, combination, and rearrangement of particles. Both the causes and methods of revealing the nature of these motions lay outside the realm of physics as it was defined within Boerhaave's curriculum. Whereas physics examined the general properties and motions of bodies, chemical effects were specific and inherent to each species of substance. Moreover, many of these effects remained latent until revealed by the proper circumstances. Only operations traditionally associated with chemistry—distillation, sublimation, fermentation, extraction, and so forth—revealed these specific properties. Thus, in addition to the "physical" properties of matter, medical students also had to understand matter as types of chemical species—spirits, oils, salts, earths—and to know the properties of each species.[56]

Chemistry also provided, in Boerhaave's words, a "method of investigating" bodies to determine their chemical properties. Students had to be familiar with the techniques of chemical analysis, and Boerhaave suggested two that were particularly useful for medicine: analysis by fire and by "effervescence."[57] Both of these methods subject a test substance to manipulation by chemical instruments, by fire and by menstrua, respectively. Analysis by fire referred to the traditional distillation of substances into their constituent parts. Boerhaave described this process in detail in his yearly chemistry courses (as described in chap. 3). The substance to be examined (a fluid or bit plant or animal matter) was placed in a distillation retort and subjected to increasing levels of heat. The various fluid fractions were collected in the receiver and analyzed to determine what type of fluid they were. Any solid dregs left behind in the retort would be analyzed as well with solvents, typically water to dissolve any saline material. From this analysis, the chemist would infer the composition of the initial substance, although Boerhaave and many other chemists doubted the effectiveness of this method, especially on complex materials.[58] Analysis by effervescence consisted of testing a substance by bringing it into contact with other known chemical species, such as specific acids, alkalis, and salt solutions, in order to generate a noticeable reaction or "effervescence." The chemical properties of the tested substance could then be established by observing the chemical species with which the unknown substance reacted.[59]

Ultimately, chemistry was important for Boerhaave's medical system because chemical properties were important for understanding human physiology and pathology. In De Methodo, he maintained that chemistry was the only art that exhibited the "natures and powers" of our food, the body's humors, and the physician's medicaments.[60] In the medical curriculum, chemistry supplied the tools for the analysis of medical substances, which buttressed the study of physiology, pathology, and materia medica. Through these tools of analysis and the battery of chemical concepts that chemistry provided, Boerhaave hoped to purge medicine of the excesses of earlier practitioners of chemical medicine. By examining his medical courses in more detail, we can see how chemical knowledge shaped his medical practices, leading him to develop a novel form of chemical medicine that was empirically grounded but more theoretically subtle than the earlier forms found at Leiden.

THE NEW CHEMICAL MEDICINE

Boerhaave's appointment to the chair of chemistry represented the success of a new type of chemical medicine, derived from the empirical and experimental practices of his medical mentors, Anton Nuck and Charles Drélincourt. Boerhaave adopted and expanded upon their practice of using chemical analysis to define the properties of bodily fluids and theorize about their physiological functions. As we have seen, he devised his own chemical system, which he deployed in place of the more piecemeal approach of his predecessors. With conceptual and methodological tools from his chemistry, he observed important bodily substances and then characterized them by how they acted or were acted upon by menstrua, their degree of acidity or alkalinity, or the way they responded to various degrees of fire. Proceeding with caution and restraint, Boerhaave hoped to use chemical knowledge to make physiological arguments regarding the function and behavior of fluids within the body, including their role in the generation of pathological states and possible treatments.

The centrality of chemical analysis and theory for Boerhaave's account of human physiology can easily be gleaned by examining his medical textbook the *Institutiones Medicae* (first edition, 1708). This textbook was based on his version of Leiden's basic medical theory course of the same name, which he had taught beginning in 1701. The first and largest section of the textbook examined basic human physiology.[61] Here Boerhaave relied on chemical concepts to define the physiological function of bodily fluids. For example, the "cystic bile" described at the beginning of this chapter participated in digestion. To characterize its precise function in digestion, Boerhaave employed terms from his chemical theory. He described cystic bile as a saponaceous menstruum: "disposing oil to mix with water and dissolving resinous, gummy and other tenacious substances." Other characteristics of this humor could only be known through chemical analysis. He related, for instance, how the humor was not combustible unless it was first dried. This characteristic could only be known through experiment: he tested the combustibility of fluid and dry bile. He also suggested that this bile was "easily putrefying" and that, once putrefied, was "very penetrating and volatile."[62] Again, this statement probably reflected the result of another test to determine the result of "putrefaction": subjecting the fluid to a low, constant heat in an attempt to generate a chemical change similar to fermentation. The result was a "penetrating" and "volatile" substance, described using common chemical terms, which suggested that the product was caustic ("penetrating") and fume generating ("volatile"). The ability of cystic bile to putrefy suggested to Boerhaave that in its volatile state, its fumes could be absorbed by other bodily humors and travel to various parts of the body, a notion that, as we shall see, supported his model of disease.

In his medical courses, Boerhaave often deployed the results of his chemical analyses to critique the physiological claims of earlier physicians. Following the work of his mentors Nuck and Drélincourt, Boerhaave often criticized the work of Franciscus Sylvius. Whereas in his academic orations he judged the method of Sylvius's chemical medicine to be overly speculative, in his medical courses he examined Sylvius's claims regarding the chemical properties of specific humors. Most of these critiques were part of his general critique of Sylvius's approach to chemical medicine, especially his use of the acid/alkali theory as a general interpretive framework for physiology and medical practice. For example, in his *Institutiones*, Boerhaave refuted Sylvius's assertions on the properties of pancreatic juice. Sylvius had argued that this humor was acidic and aided in digestion by helping to break down aliments containing alkaline parts in their composition. Boerhaave countered this claim by using his own chemical theory to provide a more nuanced explanation for the function of this fluid. He argued that pancreatic juice functioned as a mild menstruum, which weakened the "acrimonius parts" of the chyle and diluted it in the intestines in order to "render it fitter to pass the lacteals and mix with the blood." The chemical concept deployed here was the notion that some menstrua decrease the corrosive power of their solutes, such as when the addition of water to an acidic spirit decreases the acid's corrosive strength. He suggested that pancreatic juice was neither acid nor alkali, a claim he made after an effervescence test. Rather, Boerhaave described the fluid as "limpid, almost insipid"—terms that he also used to describe pure water, suggesting the aqueous nature of the fluid.[63]

Some of the most significant chemical research Boerhaave conducted on bodily humors derived from general questions in chemistry rather than specific physiological questions. This situation reflected the development of chemistry in preceding centuries by practitioners who sought to understand the composition of certain bodily fluids, such as blood and urine, as chemical reagents to be used in chemical operations and medicaments.[64] Boerhaave's efforts to examine claims made by these "chemists" often led him into physiological questions and vice versa. His study of urine provided an excellent example of how physiological and strictly chemical problems often overlapped. Uroscopy as a diagnostic tool was a long-established part of both Galenic and non-Galenic medical practice.[65] As an offshoot of this tradition, chemists and physicians had subjected human urine to chemical analysis, through distillation and other operations, to gain insight into the formation and possible treatment of bladder stones, a common and painful malady. Van Helmont, for example, wrote at length on the composition and chemistry of urine, arguing that the salt obtained from urine possessed a nature that was different from that of known acid salts.[66] By the beginning of the eighteenth century, the "salt of urine" (i.e., urea, when pure) had become a standard substance discussed in chemical and medical textbooks as both a product of the human body and a chemical species to be used in medicaments and to make other chemical products, such as phosphorus.[67]

Boerhaave discussed salt of urine at length in the *Institutiones*. He described the salt as "saponaceous" and neither acidic nor alkaline, not "muriatic" (like sea salt) or ammoniacal (like sal ammoniac—ammonium chloride), but with its own "particular nature."[68] As in the previous examples, he presented his discussion of salt of urine to undermine Sylvius's general claim that "human salts were of an alkaline nature." He asserted that he discovered Sylvius's error in 1696, when he "demonstrated" that salt of urine did not effervesce in acid liquor (a standard test of alkalinity).[69] Boerhaave also commented that salt of urine often contained a quantity of "fixed or sea salt," which derived from the common table salt used in food. The significance of this statement, which he discussed at some length, was the context in which he first discovered this fact. As he explained in the *Institutiones*, in 1696 he "was surprised to find" that the salt of urine, mixed with spirit of niter, would make a type of aqua regia "capable of dissolving gold." He found that he could produce this menstruum even if the urine from which the salt was extracted had been putrefied for "many years" or "tortured" with a fire so intense "as to produce a phosphorus."[70] What Boerhaave did not state in the *Institutiones* was that in 1696, his main chemical interests centered on the composition and transmutation of metals. His surprising find resulted from the test of a recipe for producing aqua regia from urine. This trial was clearly related to one of his alchemical projects, most likely derived from the work of alchemist, Johannus Isaacus Hollandus (fl. 1570).[71] In attempting to understand an alchemical problem, Boerhaave pieced together a physiological claim that common salt contained in food passes out of the body in the urine.

We have seen how Boerhaave deployed chemical analysis to establish the properties and functions of the body's humors. He, like Sylvius, extended his application of chemical knowledge into pathology and treatment. In his physiological system, the humors, under the proper environmental conditions, interacted with each other and with aliments according to their chemical properties to produce healthy physiological functioning in the body. Under deleterious conditions, such as the consumption of improper foods or other stresses, chemical interactions could produce disease. Exactly how the humors could cause disease derived from Boerhaave's understanding of the body as a physiological system. In his *Institutiones Medicae*, he argued that the human body was composed of two types of solid parts—fibers and membranous vessels—and various fluids contained in the vessels. He called the systematic interaction of these various parts in the body the "animal oeconomy," a term he borrowed from seventeenth-century medical professors in Leiden who advocated a mechanistic, Cartesian-influenced understanding of the body.[72] Within this "vascular" model of the body, important fluids, such as blood, transported important substances, such as nutrients and other humors, to where they needed to go in the body-machine.[73] The proper flow of the bodily fluids denoted the health and life of the organism. Retarded fluid flow in some part of the body caused illness, and a complete halt in all fluid flow indicated death.

Boerhaave argued that most diseases were caused by either a blockage in one or more of the body's vessels or an alteration in the quantity or quality of one of the body's humors, which led to impeded fluid flow. He, in fact, devised a taxonomy of conditions, which caused the blockage of vessels and that, in turn, caused various diseases. The first "species" of blockage, called "emphraxis," was the plugging of the vessel by corrupt fluid matter or debris within the vessel. Other "species" included various inflammations or tumors in the "membranes" of the vessel, or just outside it, that constricted the vessel, encouraged abnormal growth of a vessel, or allowed the collapse of a vessel through overdistension and evacuation.[74] All of these blockages led to an inflammation of tissues as fluid and solid parts became increasingly impacted. If the humor in question was blood, as it often was, the site of the inflammation began to heat up as a result of the attrition (collision and rubbing) of particles, since the blood was still being pumped by the heart. As the heart rate sped up in an attempt to maintain blood flow, the increasing attrition of blood particles in the blood vessels generally led to a fever and, if the pressure at the site of the inflammation was great enough, pain as well.[75]

An important aspect of preventing and treating disease within Boerhaave's system was to understand how these blockages were formed. Blockages that occurred as a result of chemical action in the humors were a form of "emphraxis." Boerhaave modeled his explanation for these kinds of blockages on traditional, Galenic medicine. The ancient Greek physician Galen (fl. ca. 150–180) famously posited that illnesses could arise though the "concoction" or "putrefaction" of one of the four traditional humors (blood, phlegm, yellow bile, black bile). This putrefaction generated heat, which became a fever as it was conveyed to the heart and then to the rest of the body in the form of either a "pneuma" in the arteries (which Galen believed contained no blood) or a "sooty vapor," conveyed by the blood in the veins.[76] Boerhaave, like most of his contemporaries, had discarded Galen's four-humor system as overly simplistic, preferring to identify and define the body's numerous humors through physiological experimentation and chemical analysis. Nevertheless, the framework of the humoral theory, which depicted the body as a system of moving fluids, remained the model of human physiology and acted as a theoretical tool to define the causes of health and disease for most physicians during the seventeenth and eighteenth centuries. Even as novel physiological discoveries, such as the circulation of the blood, or new theoretical approaches, such as Paracelsianism or "mechanical medicine," were established, physicians adapted the humoral model to fit their new circumstances. Paracelsus, for example, devised the idea of "tartaric diseases," in which illnesses, such as consumption (tuberculosis of the lung) or gout, were caused by abnormal accretions ("tartar") generated by chemical concoction and which prevented normal physiological functioning.[77]

One traditional, Galenic cause of disease found in Boerhaave's pathology was the idea that improper dietary habits caused corruptions in the body's humors. In Boerhaave's theory the types of corruptions that a specific type of food could cause were based in large part on the food's behavior as a chemical species outside the body. In vitro, some foods, like most vegetables and fruits, fermented into acidic substances, whereas other foods, such as meat and fish, could putrefy to form alkalis.[78] When a person suffered from improper digestion, caused by conditions ranging from improper "animal motion" in the viscera to simply overeating, these improperly digested foods, which Boerhaave termed "crudities," began to ferment or putrefy. When they were absorbed into the blood, they could further putrefy according to their natural inclinations into acidic or alkaline "acridities." These acridities could then cause disease symptoms by themselves. Boerhaave often determined what these symptoms might be by observing the chemical behavior of bodily tissues exposed to various acids, alkalis, and other salts in the laboratory. For example, an excess of "acidity" in the blood could lead to itches and pimples, but eventually to a more serious deterioration of the body's tissues, especially bones and teeth. Boerhaave arrived at this conclusion based on an experiment performed by Frederik Ruysch, the famed Amsterdam anatomist, who soaked bones in acid and reported that they became soft and flexible.[79]

Serious conditions could develop when the "acridities" combined with other components in the blood to form insoluble masses that blocked the capillaries (i.e., emphraxis). Boerhaave argued that these types of blockages usually formed through the combination of alkalis and "earths." Again, he based this conclusion on the knowledge of the action of chemical species outside the body, in this case the "intimate" union of an alkali with an earth through the action of fire to form an insoluble "glass" or chalk. He also had some evidence from within the body as well. On the basis of the results of chemical analysis, he knew that the stones that formed in the bladder, kidneys, and gall bladder were insoluble bodies composed of an alkali and an earth.[80] From these facts, he devised a theory of stone formation. The "earth" component of the stone came from the body itself, as particles of earth were sloughed off from the solid parts of the body through the normal motion of tissues and fluids. In bladder and kidney stones, the alkali component of the stone came from the alkaline component of putrefying urine; in gall stones, from an alkali generated by putrefying bile.[81]

Boerhaave applied the alkali/earth model of stone formation in his general theory of disease. Alkalis that entered the blood stream, usually as a result of improper digestion, could combine with sloughed particles of earth present in the blood to form blockages in the circulatory system. Blockages of this type, which hindered flow through the capillaries, caused inflammations and fever.[82] We have already seen an example of this type of pathology. Recall Boerhaave's description of "cystic" bile at the start of this chapter. Its function in digestion was to dissolve resinous, gummy, and "tenacious" substances, but it was easily "putrefied." When this happened, it formed a penetrating and volatile alkali. Just like a stone could form in the gall bladder, vapors from the putrefied bile, absorbed by the lacteals with digested food, reacted with "earths" found in the blood to form blockages.[83]

One might wonder how Boerhaave could justify his acerbic criticisms of Franciscus Sylvius's chemical medicine, given the role that chemical properties of acids, alkalis, and other chemical species played in his theories of physiology and pathology. He clearly accepted the premise that some diseased states were caused directly or indirectly by acidic or alkaline substances in the body. Like Sylvius, Boerhaave even extended this premise to treatment; patients who suffered from an excessive acidity, for example, were treated with alkali powders, such as "crab's eyes," and a diet of foods, such as meat and fish, that became alkaline during digestion.[84] Ultimately, his criticism of Sylvius stemmed from the methodological shortcomings he perceived in Sylvius's system. He accused Sylvius of privileging his theory over the empirical evidence, leading to a misinterpretation of that evidence. He pointed out Sylvius's mistakes on several occasions in his works, as shown by the examples in this chapter (e.g., cystic bile and pancreatic juice). In most of these cases, Sylvius assigned an acidic or alkaline nature to a bodily fluid, which Boerhaave later tested only to find that it did not pass a simple effervescence test. Because of this, he argued that all healthy bodily humors were neutral or "bland." Only through physiological processes, both healthy (digestion or "concoction") and pathological (putrefaction), did they become acidic or alkaline. Those aliments and humors, such as cystic bile, that had a tendency to become alkaline he classified as "alkalescent," whereas those that had a tendency to become acidic were "acescent."[85] Thus, Boerhaave's criticism of Sylvius was one of methodology and the degree to which theory can reasonably be applied to specific circumstances, not one of fundamentals. For Boerhaave, the body was much more complex than its depiction in Sylvius's system and therefore demanded a more nuanced theory.

In his lectures and textbooks, Boerhaave stressed the limits of chemistry and the other arts and sciences to describe the workings of the human body completely. Take, for example, Boerhaave's use of chemical analysis in his attempt to understand the physiological functions of the blood. In a private medical course he gave in 1717 titled De Legibus, Secundum Quis Humores in Corpore Humano Sano Moventur, he included a section in which he discussed his analysis of blood serum through distillation. Ultimately, he was able to recover eight separate constituents, including several humors, salts, and one earth. This result verified his contention that blood was "complex" but also alluded to the function of blood in the animal oecomony. The blood, carried throughout the body by its circulation, had the job of transporting all of the body's other diverse humors, salts, and earths from place to place. In effect, blood was a very complex menstruum, having to dissolve substances of various types within itself and still remain homogeneous.[86] Boerhaave, however, was skeptical regarding the ability of the chemical arts to provide him with enough information to make solid physiological claims in this case. He remarked to his students on the heterogeneous nature of the components of blood produced by chemical distillation, suggesting that one might doubt that these components were present unaltered in the original serum. This remark derived from his skepticism about analysis by fire, which was a central tenet of his chemistry. In the end he concluded that the "action of the blood in the body" could not as yet be determined by chemical methods.[87]

Employing acerbic rhetoric like that found in his oration *De Chemia suos Errores Expurgante*, Boerhaave was able to vanquish rival medical systems, like Sylvian chemical medicine, while adopting some of their basic principles into his own work. He, however, saw his own chemical medicine as programmatic rather than dogmatic. He did not present to his students a complete set of medical precepts that only needed to be applied appropriately. Medicine, like chemistry, was in the process of being purged of its errors, and many of its theoretical precepts had not been established. Within the pedagogical context in which Boerhaave worked, he saw the gaps in his system and the limitations of his methods as opportunities for further research, especially for his students. When he addressed topics in his courses or textbooks that had not yet been placed on a sufficiently sound footing, he recommended to his students how to proceed methodologically to solve the problem. For example, in the *Elementa Chemiae* he described the distillation of human urine and the various fractions that the astute chemist recovered from this operation. He asked his readers: "what does this get you (as a physician)?" His answer was "very little." As in his distillation of blood serum, chemical methods in this case did not provide enough reliable data to connect the results of analysis to physiological states in the body. So, Boerhaave recommended that his students continue to read the "ancient works" on urine and to conduct further chemical analysis. By diligently following this method, he assured them, the art of uroscopy may someday be "reduced to regular laws."[88]

CONCLUSION

Boerhaave asserted in his *Elementa Chemiae* that "the Art of Medicine . . . treats the Human Body and the powers and effects of other bodies upon it, neither of which can be thoroughly understood without the assistance of chemistry."[89] Chemistry furnished the physician with a battery of analytical tools and theoretical concepts with which to characterize the "chemical" properties of bodily substances, aliments, and medicaments in order to make sense of their actions within the body and to determine proper treatments and prophylaxis. In this, Boerhaave continued an established tradition among Leiden's anatomical and physiological experimentalists, including the illustrious Franciscus Sylvius. To achieve the level of integration of chemistry and physiology that he wished, however, he had to distance his chemical medicine rhetorically from that of Sylvius and other chemical physicians, such as Joan Baptista van Helmont, despite that fact that he incorporated may of their core ideas into his medicine. In the end, Boerhaave produced a new chemical medicine that he could present as empirical and disciplined, and as one that followed a "Hippocratic" methodology.

The success of Boerhaave's chemical medicine ultimately shaped the development of chemistry as a scientific discipline. Because it provided an important theoretical framework and research method for his medicine, his chemistry became institutionalized within the Leiden medical curriculum. Medical students needed chemical knowledge to understand human physiology and Boerhaave's theories of pathology and treatment. This situation justified the inclusion of chemistry in the medical curriculum by connecting the field to institutional practices—physiology lectures, anatomical demonstrations, the *praxis medica* course—that were well established in the Leiden medical faculty. As Boerhaave's students began to transplant the "Leiden model" of medical education to other medical faculties, Boerhaave's theories of physiology and pathology, and the chemical knowledge that served as their foundation, spread throughout Europe.[90]

CHAPTER FIVE

Instruments and the Experimental Method

When Jacob le Mort, the professor of chemistry at Leiden, died on March 1, 1718, Herman Boerhaave was the unquestioned choice to fill his chair. At their meeting on June 24, the curators and burgemeesters appointed Boerhaave professor of chemistry, stating that he had given "colleges" in the subject "for several years with much success." For his new responsibilities, they increased his salary by 200 guilders.[1] In fact, Boerhaave had continued to teach his chemistry lecture and demonstration courses regularly since 1703. His private lectures, which students paid to attend, typically attracted many more auditors than the free lectures given by Le Mort. This situation had only served to reinforce Le Mort's lackadaisical attitude regarding his teaching duties, and during many winter terms in Leiden, Boerhaave's was the only chemistry course that interested students could attend.[2] By 1718 he was already de facto fulfilling the curricular duties of the professor of chemistry.

As professor of chemistry, Boerhaave fulfilled his obligation to teach at least one public course (i.e., open to all matriculated students) in chemistry by offering an extended series of lectures on the chemical instruments. The instruments course redefined the pedagogy of chemistry in Leiden by examining the nature of the chemical instruments—fire, air, water, earth, and chemical menstrua—though demonstration experiments. Previous lecturers, such as Le Mort and Carel de Maets, had structured their courses following the didactic model, in which the theory of chemistry was presented briefly and then the lecturer moved on to show operations and recipes. Boerhaave presented these kinds of procedures in his operations course, which he taught regularly beginning in 1703 and continuing until 1728.

In the instruments course, however, he performed experimental demonstrations that he had designed and routinized to elucidate theoretical principles. These experiments did not demonstrate how to make anything but rather exhibited the properties of matter and how the instruments affected them. Boerhaave's immediate model for this type of demonstration was the experimental physics course offered by the natural philosophers in the Leiden arts faculty. While Boerhaave was a student, Wolfred Senguerd and Burchard de Volder had given these courses in Leiden's *theatrum physicum*, and in 1718 the "Newtonian" physicist Willem Jacob 'sGravesande (1688–1742) continued this practice. The pedagogical aim of these courses was to demonstrate the theoretical precepts that had been established in the physics lecture course. Thus, the demonstrations were intended to provide paradigmatic, empirical examples of established principles. In Boerhaave's instruments course, the demonstrations performed the same function, except that he conflated the typical division between the *theoria* and *practica*. The empirical demonstration of a phenomenon and its theoretical discussion happened simultaneously. In effect the instruments course brought Boerhaave's empirical philosophy to the foreground of his chemistry. Chemical theory was to be determined by sense perception and experimental demonstration, not the other way around.

The main pedagogical problem for the physics demonstrators, and for Boerhaave in the instruments course, was the managing of sensory data. Demonstrations had to guide the student observer to see the demonstrated phenomenon in the correct manner so that he would arrive at its correct interpretation. Both Boerhaave and the Leiden physicists designed their experiments to do this, often creating new techniques and novel apparatus specifically for use in their demonstration courses.[3] Even with these aids, students still had to be guided to see the phenomenon in the proper manner. Thus, professors embedded their empirical demonstrations in explanatory structures designed to connect the proper terms and concepts to what the student appended through the senses. Boerhaave managed his demonstrations by embedding them within a literary structure, which I call the "demonstrational" form, that pointed out, defined, and explained the relevant aspects of the phenomenon under display.

The instruments course, however, went beyond the aims of experimental physics in Leiden. Boerhaave also designed his course to show the manner of *generating* theoretical principles from experiment. His course provided a model for practicing chemistry as a form of experimental philosophy. At the onset of the course, Boerhaave outlined a "method" for generating chemical knowledge though structured experimentation following the tenets of Francis Bacon. He employed "Baconian" principles and methodologies—the construction of ordered experimental histories, reasoning from empirical observation to argue for unseen causes, and the emphasis on practical utility—to structure much of his chemical work.[4] There are two different uses of the term "method" at work here: the "method of discovery" and pedagogical method. The first usage derived from late scholastic philosophy and signified a set of resolutive and compositive techniques used to generate novel (logical) propositions for the purpose of establishing causes for known effects. This type of method was refined and reinterpreted in the seventeenth century by philosophers such as Descartes and Bacon to resemble what moderns call a (or *the*) scientific method, but which for clarity's sake I refer to as a "methodology." The second type of method, also known as *ordo*, concerned the proper ordering of topics in a discipline for pedagogical efficiency. Pedagogical method reduced a given discipline (or art) to its precepts or principles and presented them so that each topic provided the student with a foundation of knowledge and skills needed to comprehend the later topics.[5] In the instruments course, Boerhaave conflated these two notions of method. The pedagogical *ordo* of the instruments course also displayed a method of discovery—a series of logical and empirical proofs for the claims he was demonstrating.

The process of composing and presenting his instruments course connected Boerhaave's essentially academic approach to chemistry to that of other experimentalists and public lecturers outside of the university. By the early eighteenth century, public lectures and demonstrations in experimental physics, chemistry, and anatomy were popular forms of entertainment and also important vehicles for public education in the sciences. These public lecturers faced the same problems as university lecturers, like Boerhaave, in conveying the content of their courses efficiently and working out the demonstrations displayed during their courses beforehand. Recent work on public lecturers has argued, in fact, that these demonstrations constituted an important form of natural philosophy in their own right.[6] Like the public demonstrators, Boerhaave had to solve the material problem of making his experiments perform in the correct fashion during the lecture demonstration. He routinized the performance of his experiments and manipulated equipment to amplify the phenomenon he wanted to display to his audience. This process was similar to the process of "trying" an experiment as described by experimentalists at the Royal Society of London and other such locations.[7] In addition, the instruments that Boerhaave deployed in his lectures were often specially made for this purpose. To procure this specialized equipment, Boerhaave collaborated with instrument makers such as Samuel van Musschenbroek (for his air pump) and Daniel Gabriel Fahrenheit (for thermometers), who also supplied their wares to other demonstrators. Fahrenheit, in addition to making instruments, was himself a public science lecturer.[8] Besides procuring instruments, Boerhaave used Fahrenheit's experimental knowledge to shape his own lectures on the instrument "fire." Ultimately, the act of designing, organizing, and arranging demonstration experiments brought to light new phenomena and forced Boerhaave to refine earlier claims and precepts in his chemistry.

This chapter examines Boerhaave's instruments course both in terms of its pedagogical structure and in terms of the work of designing and trying demonstration experiments. I examine Boerhaave's Baconian "method of discovery," which he elaborated at the start of the instruments course, and discuss how the pedagogical structure of the course provided a model of practice for following this method. I also look at how he devised his new "demonstrational" pedagogical method to manage sensory perceptions from the demonstrations themselves. Finally, I survey the process by which he put together his instruments course, both by disciplining his experiments and integrating them into his course. By focusing on Boerhaave's work with the thermometer and the air pump, I recount the ways in which the act of composing his course generated novel theoretical and practical approaches to chemistry. Following recent work on public science demonstrations, I show that Boerhaave's pedagogical work for his instrument course was a form of natural philosophy.

BUILDING A PHILOSOPHY FOR CHEMISTRY

During the three months following his appointment as professor of chemistry, Boerhaave composed and delivered his academic oration *De Chemia suos Errores Expurgante* and the first few lectures of a new, public course on the chemical instruments. Unlike his approach to the instruments in his original lecture course, his approach in the new instruments course featured demonstration experiments as its backbone. In addition, the course was an extended affair: he did not conclude the material in five months, like he did in his other chemistry courses. Rather, he presented his examination of the instruments as one long course, with new material being covered each term from October 1718 through May 1728.[9] In effect, the instruments course was similar to some of the private medical courses that Boerhaave had presented on specific topics. These courses were one-time affairs, and some lasted several years.[10] During his ten-year professorship of chemistry, he gave three chemical courses each winter term: his two private (i.e., subscription) courses—the lecture course on chemical *theoria* and the operations course—and one public, ongoing course on the chemical instruments.

In the first five lectures of the instruments course, Boerhaave defined the proper domain and methodology for chemistry. Early on, he pointed out that he had described this proper methodology in an oration he had had given in 1715, *De Comparando Certo in Physicis*, but that he would review the relevant points again so his auditors might discern their use in chemistry.[11] In the oration, he equated his approach to natural philosophy with that of Francis Bacon. He asserted that Bacon was "the greatest by far in investigating all subjects that can be attained by human science" and that "we are indebted to the warnings, precepts, and experiments of this man."[12] As was well recognized, Bacon argued that the "axioms" of natural philosophy must be grounded upon sensory observations. These observations must be collected and organized into natural or experimental histories, from which philosophers established general propositions through induction.[13] Boerhaave claimed that all valid progress in natural philosophy (and medicine) derived from this method. In fact, the "Baconian" method that he described in *De Comparando* was virtually identical to the "Hippocratic" method for medicine that he had described in his earliest orations.[14] When he deployed this method within a medical context, he called it "Hippocratic," but within the context of physics it was "Baconian."[15] Just as he cast as Hippocratic the work of modern physicians whom he admired, Boerhaave depicted the innovators in natural philosophy as following the Baconian method. In *De Comparando*, for example, he argued that Isaac Newton arrived at the idea of mutual, gravitational attraction between bodies by collecting numerous observations and, only after careful consideration, positing the existence of this general principle.[16]

At the start of the instruments course, Boerhaave advocated for Bacon's methodology of making experimental histories and arriving at general principles through induction as the methodological basis for chemistry. He considered chemistry to have the same methodology as physics, and the bulk of the these introductory lectures concerned the demarcation of chemistry from the mixed mathematical sciences (i.e., mechanics, hydrostatics). In this he continued arguments that he first presented in his course De Methodo Addiscendae Medicinae.[17] The heart of the demarcation problem, as Boerhaave constructed it, was that most chemical phenomena were not reducible to mathematical laws. He emphasized, however, that this deviation from mathematical law was not the result of any flaw in the methodology of chemistry but rather reflected a fact of nature: mathematics was simply not applicable to all phenomena. He defined bodies (the objects of study in physics) as those entities that had extension, were impenetrable, mobile, capable of rest, divisible into similar masses, and capable of forming various (other) masses. These bodies had properties that the physicist was capable of comprehending through sense data. Among these properties, some were common to all bodies, and some were proper only to singular bodies. Boerhaave argued that these common properties were the proper domain of the mixed mathematical fields. The common properties included extension, solidity, weight, figure, divisibility, motion and the "unity of certain cohering elements." These general properties were explicable through mathematics, especially geometry, which demonstrated and devised the most accurate representations of these properties. Through the use of geometry, mechanics "by the same clarity and faith," had investigated more complex phenomena by employing the mathematical representations for motion, gravity, mass, and fluid motion.[18]

Boerhaave asserted that chemistry illuminated the second kind of property: those proper only to singular types of bodies. He warned that mathematical practitioners "fall into error if they suppose that all nature may be drawn out in this way [i.e., mathematically], as if beyond these general [properties], others are not in bodies and in fact equally effective."[19] This error was similar in form to the ones that Boerhaave described in his orations in which medical practitioners universalized claims that were properly applied only in specific circumstances. In the instruments course, he reasserted that the Cartesian "sect" was guilty of this error by trying to explain all knowledge of nature through "universal demonstrations" that only acknowledged the general, mathematical properties of bodies. Citing Bacon, he explained that the "singular properties of movement" in bodies were latent, and thus they could not be deduced from universal principles but were only known through observation by the senses. These singular properties that were beyond the universal demonstrations of mathematics were the proper domain of chemistry. To emphasize this point, he listed several examples in which these latent properties became manifest. He cited the reactions of acids and alkalis, of acids and metals, of acids and oils, of antimony metal and gold, and of sea salt and sulfur.[20] Each of these examples identified a phenomenon that arose only when two different types of body were brought into contact but that could not be predicted based on the mathematical properties of those bodies.

After differentiating the domain of chemistry from that of the mathematical fields, Boerhaave outlined a method by which chemistry would reveal the latent properties of bodies. He stated that bodies were first known through chemistry by resolving them into their "most simple elements" and then joining those simple parts together again. Then, bodies were to be "agitated by various grades of motion." The chemist must gradually collect more and more instances of bodies being resolved or brought into motion; in effect, the chemist must construct an experimental history of chemical species. Boerhaave urged that during each of these processes, the chemical practitioner must "diligently observe and note all phenomena" that appeared. Eventually, from all of these observations, the chemist will "elicit" (*elicio*) general principles, which take the form of rules. Through this process, he added, "the vast utility and absolute necessity of chemistry is understood."[21]

DEMONSTRATION AND METHOD

Boerhaave designed his instruments course to examine the properties of the chemical instruments, the theoretical core of his system of chemistry. The course itself was a mix of lectures and experimental demonstrations, which were "methodized" for pedagogical efficiency. Unlike his previous courses, however, in the instruments course Boerhaave wrestled with the problem of linking empirical information (the student's sensory perceptions) with the claims and concepts of his chemical theory. To manage this empirical information, Boerhaave devised for his course a new "demonstrational" structure, which he used to organize his lecture notes and presentations to students. This new pedagogical method was designed to guide students into interpreting phenomena according to the framework of Boerhaave's instrument theory and acculturate them into the experimental philosophy that he used to establish his theory.

Boerhaave's approach to presenting experimental demonstrations derived from the innovations in pedagogical method developed by the Leiden physicists in the arts faculty. Because, as Boerhaave had argued in the introductory lectures of his instruments course, chemistry had the same methodology as physics, physics was the natural place to look for pedagogical models. The Leiden natural philosophers in the arts faculty spoke of "method" as the systematizing and simplifying of knowledge for the purpose of conveying this knowledge to students easily and efficiently.[22] At the beginning of the seventeenth century, "method" signified the ordering of *questiones* in scholastic philosophy lectures, but by the end of the century the Leiden philosophers had developed alternative structures and methods of presentation. Most striking was Burchard de Volder's experimental physics demonstrations, which he performed in the university's *theatrum physicum* beginning in 1675. De Volder made the pedagogical intent of his experimental demonstrations clear. In his petition to the Leiden curators asking for permission and resources to found the *theatrum,* he asserted that his demonstrations would serve to confirm the principles already established in the *physica theoretica* course. Thus, the point of De Volder's experimental demonstrations was to provide examples of phenomena as a means of reinforcing and clarifying theoretical principles for students. Within this controlled context of demonstration, empirical evidence was a valid pedagogical tool for helping students to understand theoretical principles, but De Volder did not offer such experiments as a means of discovering new knowledge. For De Volder, reason, not experience, was still the chief guide to discovery.[23]

Boerhaave incorporated the form and ideology of De Volder's pedagogical demonstrations into his instruments course. Both De Volder and Boerhaave followed the traditional tenet from scholastic pedagogy that true knowledge was knowledge of universals. As Peter Dear has shown for seventeenth-century pedagogical texts, empirical phenomena or "experiments" were almost always portrayed as universal *experiences.* These experiences were not discrete historical events or records of individual observations—how nature had *behaved* in a given circumstance—but rather were general statements of how nature *behaves* in a given circumstance.[24] Boerhaave's demonstration experiments functioned in this same way—as a means of creating universal experiences for his auditors. Although his demonstrations were discrete events, he construed them as universal representations of natural principles. The pedagogical point of a demonstration was not to test the principles under discussion but was to educate the students' senses and discipline their minds to recognize the relevant phenomena in the demonstration and connect their perceptions to the technical language and concepts that grounded Boerhaave's natural philosophy.[25] Even though there were seemingly vast epistemological differences between Boerhaave and his mentor—De Volder argued for the primacy of reason, whereas Boerhaave maintained that the senses were the basis for all natural knowledge—within a pedagogical context such differences were moot. The aim of pedagogical method was not to debate philosophical subtleties but to present the principles of a discipline to novices in an unambiguous manner. Thus, Boerhaave, like De Volder, worked out his chemical demonstrations in the laboratory before the lecture and then embedded them in a methodized structure of precepts designed to train students to see the phenomena he demonstrated as empirical examples of theoretical principles.

Boerhaave was mindful of the problems involved in conveying information though experimental demonstrations, so he adopted a new pedagogical structure, the demonstrational method, to guide his students toward his interpretation of the demonstrations. As he recorded them in his lecture notes, each demonstration consisted of an *experimentum*, which described the procedures and results of the demonstration experiment being performed. This description was followed by a series of numbered *corollaria*, which listed the inferences and implications to be grasped from the experiment. This structure mirrored that of De Volder's experimental physics course and also followed closely the method by which Isaac Newton presented experiments in the *Opticks*.[26] The goal in each of these texts was to guide the reader or student, if the demonstration was presented by a lecturer, toward the proper inferences and conclusions about them. For example, in Boerhaave's first *experimentum* in the instruments course, he exhibited a simple iron bar that expanded in length when placed in a flame. From this simple demonstration, he presented several generalizations as *corollaria*. He asserted that all hard, solid bodies expanded in this manner when exposed to "fire" (the instrument, not necessarily flame). He argued that a greater amount of heat increased the expansion proportionally. Finally, he reported that "internal motion" could be induced in all parts of hard bodies in this way.[27] These inferences do not necessarily or readily follow from the experiment itself, but rather they served to alert Boerhaave's auditors what the significant phenomena and claims that could be derived from them were. In the example just given, the student was directed to see expansion as the significant phenomenon, rather than, say, the glowing of the bar, its change in color, or the sensation of the heat that it emitted. Thus, Boerhaave's demonstrational method, as a literary form, was an organizational tool for his lecture notes and, as a form of experimental practice, was a means of drawing the student's attention to the relevant aspects of the displayed phenomenon.

In practice, Boerhaave embedded his demonstration experiments within a methodized structure that combined both theses and experimental demonstrations. Single demonstration experiments did not stand on their own, but rather Boerhaave constructed chains of demonstrations designed to elucidate a theoretical conclusion by exhibiting the behavior of a phenomenon though variations in circumstance. He enumerated the general conclusions in the *corollaria* of the individual demonstration experiments or, occasionally, in a scholium at the end of the series. He often provided context for these chains of demonstrations by defining experimental problems through numbered theses. As I explained in chapter 3, each thesis elaborated a definition, theoretical principle, or observed fact. A set of theses taken as a whole established the important pedagogical precepts for any topic in the course. Expanding on the "instruments" section of his earlier chemical lecture course, Boerhaave applied the thesis method to present information that he could not demonstrate experimentally, such as definitions, logical arguments, and the observations and claims of other philosophers. The number of demonstration experiments for each chemical instrument varied widely. For example, Boerhaave's lectures on fire were almost completely composed of demonstration experiments, the lectures on air and water saw the first half of the topic elaborated by theses and the second half by experiments, but the lectures on earth were entirely composed of theses. In effect, the extent to which he employed either the thesis or demonstrational method depended on the vagaries of the topic and on the material constraints of the lecture space.

To understand how Boerhaave's hybrid method worked together, let us examine the first series of experiments in the instruments course. "Fire" was the first chemical instrument that he surveyed. He began his discussion of fire by relating the general nature of fire and describing the philosophical problems that its study entailed. This first section was structured as numbered theses, and its purpose was to set up an experimental problem that he could then (seemingly) solve though experiments. The first thesis briefly recounted Boerhaave's assumption about what fire was: a subtle fluid that mixed with other bodies and induced motion in them. The subtle and elusive nature of fire, however, created a problem for its study, because it was difficult to apprehend by the senses. Boerhaave argued in the next three theses that one must find a sensible property or a "sign of fire" that was present in all phenomena involving fire and by which one could demonstrate the presence of fire experimentally. He presented a list of empirical qualities that were traditionally associated with fire and were therefore candidates for being such a sign: heat, light, the rarefaction of fluids and solids, combustion, fusion, and so forth. He rejected light, combustion, and fusion as possibilities on the grounds that they were not present in all phenomena related to fire. Heat (*calor*) posed an interesting problem, because it was a direct, sensible result of the action of fire. But Boerhaave pointed out that our direct sensation of heat was unreliable, and he supported this claim by relating a general experience from the reports of persons who explored underground caves. In winter, the cave felt warm, but in summer the cave felt cool, despite the fact that the cave remained at the same temperature regardless of the season. So he argued that the only reliable "sign of fire" was the rarefaction of solids and fluids. This "effect" would not fool the senses like the direct sensation of heat, was observable in all phenomena involving fire, and most importantly, enabled one to measure the increments and decrements of the fire present in bodies.[28]

Having established with theses that the expansion of bodies was the "sign of fire," Boerhaave began a series of experiments to demonstrate this expansion and establish general inferences from the observation of this phenomenon. We are already familiar with the first experiment: the expansion of an iron bar when placed in a flame. The second *experimentum* merely consisted of observing the iron bar return to its original size as it cooled in the air. Boerhaave appended thirteen *corollaria* to this experiment, which generalized for both cooling and the heating the inference from the first experiment—that the expansion was proportional to the amount of fire in the body. Ultimately, he asserted that "heat" and "cold" imparted to (or took away from) solid bodies motion that was distributed to all the parts of that body.[29] In the next two *experimenta,* Boerhaave demonstrated the expansion of air, and for these demonstrations, he employed a J-shaped "Dutch"- or "Drebbel"-type air thermometer. This instrument consisted of a J-shaped glass tube with the high end open to the air and a sealed bulb on the low end. The tube was filled with water such that a quantity of air remained in the bulb and the water level was discernible in the long neck of the tube. By breathing on the bulb and holding it in his hands, Boerhaave was able to show the expansion of the air from the heat of his body based on the change in the liquid level in the tube. He then showed its contraction as he removed his hands. Again, he generalized from these demonstrations, stating that fire incrementally expanded the air based on the amount of fire present.[30]

The next four experiments concerned the expansion of liquids as the result of fire. Each experiment centered on a thermometer that used a different thermometric liquid. From these demonstrations, Boerhaave detailed the measured expansion of each liquid over a range of temperatures. In the second experiment of this set, he displayed a spirit of wine thermometer that had been constructed by Daniel Gabriel Fahrenheit (1686–1736) (see fig. 4). In the *corollaria,* Boerhaave described this thermometer to his auditors in terms of the volume of spirit contained in the bulb and the proportion of this volume that the spirit expanded when heated. He stated that the bulb contained 1,933 parts of spirit and the tube 96 parts, one for each degree according to the scale marked on the side of the tube. He explained

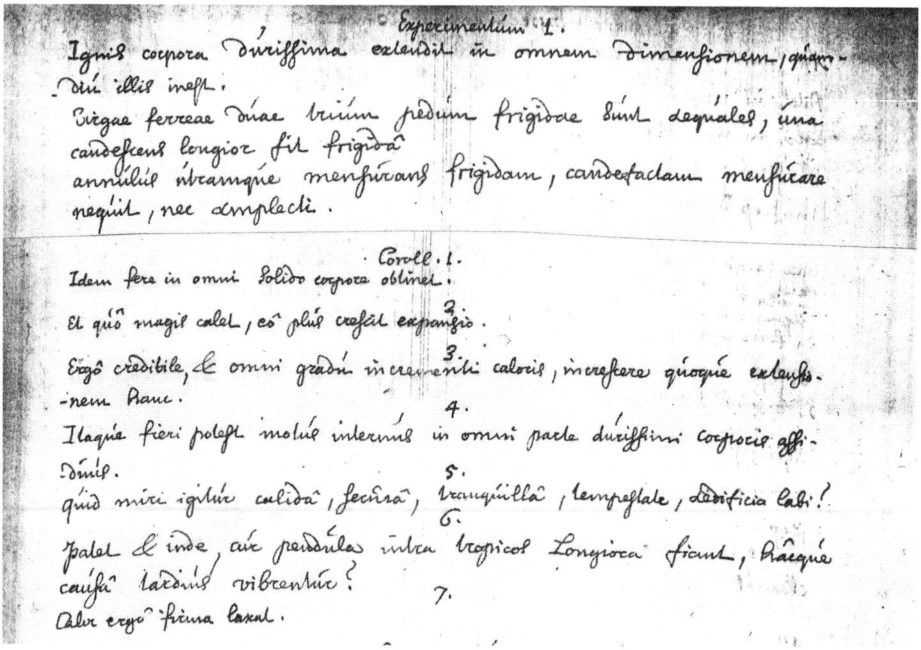
Experimentum 1.
Ignis corpora durissima extendit in omnem dimensionem, quamdiu illis inest.
Virgae ferreae duae trium pedum frigidae sunt aequales, una candefacta longior fit frigidâ
annulus utramque mensurans frigidam, candefactam mensurare nequit, nec amplecti.
Coroll. 1.
Idem fere in omni solido corpore obtinet.
2.
Et quo magis calet, eo plus crescit expansio.
3.
Ergo credibile, et omni gradu incrementi caloris, increscere quoque extensionem hanc.
4.
Itaque fieri potest motus internus in omni parte durissimi corporis assidue.
5.
quid mirum igitur calidâ, serenâ, tranquillâ, tempestate, aedificia labi!
6.
patet et inde, cur pendula intra tropicos longiora fiant, hacque causa tardius vibrentur?
7.
Calor ergo firma laxat.

FIGURE 3. Experimentum 1 from the instruments course. Source: VMA, MS 7, 3r.

that the spirit expanded by approximately 1/20 of its volume (i.e., 96/1,933 parts) when warmed from the "greatest cold" (0°F) to the "greatest heat" (96°F) on the scale. Boerhaave noted that the spirit boiled at 174°, at which point it had expanded by 1/11 of its volume.[31] He then displayed thermometer tubes filled with oil of turpentine and "pure rainwater," noting the expansion of each from room temperature (about 52°F) to the point at which the water boiled (212°F). Finally, Boerhaave displayed one of Fahrenheit's quicksilver thermometers. Following the format that he established with the spirit thermometer, he reported that the bulb contained 11,520 parts of mercury and that the mercury expanded 1/28 of its volume between 0° and 212°.[32]

The expansion of bodies caused by the presence of "fire" was only the first of its properties that Boerhaave presented through a series of demonstrations. In a scholium at the conclusion of the thermometer demonstrations, he summarized what he had accomplished and provided a link between these first experiments and those of the next series. He related that fire had the power of rarefying (*rarefaciendi*) all bodies, and he further asserted that "this property" was "always present" and "everywhere."[33] Boerhaave next endeavored to illustrate fire's omnipresence through demonstration and appeals to common experience. In demonstration, he rubbed two pieces of iron together until they became hot through "attrition" (*attrita*). This phenomenon, he explained, was observable in all bodies, at all times, and in all places regardless of the ambient temperature and therefore proved the ever-present and expansive nature of fire. In the final *corollarium* of the demonstration experiment, he alluded that this demonstration (in conjunction with the earlier expansion experiments) also indicated that the observable effects of fire (i.e., heat and expansion) were caused by motion.[34] In this manner, Boerhaave logically linked chains of demonstrations and related inferences to establish the properties of fire. Each series of demonstrations established some property of fire and proposed a new property to be investigated by the next series. The definition of the instrument, fire, was the sum of these experimentally determined properties.

Within the demonstrational method, we can see Boerhaave's empirical natural philosophy at work. The demonstration of the properties of fire, as the subtle fluid that caused the phenomena of heat, *proved* its existence. Boerhaave's reasoning followed a form of philosophical induction that he outlined in *De Comparando Certo in Physicis*. Here he argued that one could infer the existence of unobservable entities if numerous empirical observations pointed to their existence. Following a line of arguments first suggested by Francis Bacon, he asserted that "atoms" were unobservable, but by collecting empirical observations of changes in matter, one can reasonably infer their existence.[35] This same line of reasoning underlay Boerhaave's demonstrations on fire. In the process of exhibiting how heated substances expanded and, later, how heat was generated through attrition and combustion, he was building the case for the existence of his subtle fiery fluid. Many of the inferences that he made from his demonstrations in the *corollaria*, however, deviated from the central argument. For example, in one of the *corollaria* from the expanding-bar *experimentum*, Boerhaave suggested that the phenomenon of expansion caused by fire was responsible for the slower oscillations of pendulums in tropical regions.[36] This type of statement should be interpreted as a suggestion for further application of theory. In reality, Boerhaave's audience was not composed of philosophers whom he needed to convince of his assertions, but rather of students whom he was training to interpret and manipulate phenomena in the way that he did. His aim was to show the potential applications of his theory as well as prove its validity. As a point of pedagogical practice, he defined fire as a subtle fluid at the outset, and the rest of his lectures on fire constituted a prolonged and structured definition of the nature and properties of this fluid. The effect, and indeed the guiding principle, of this approach was that all claims, regardless of how novel, were presented as established knowledge. What Boerhaave exhibited to his students, in addition to the content of his claims, was a method of demonstrating the validity of his claims from empirical observations.

That Boerhaave's experimental demonstrations were intended to reinforce and display established knowledge claims was exemplified in other parts of his instrument course. Whereas Boerhaave's lectures on fire were primarily composed of demonstrations, his lectures on the other four instruments were not. In his lectures on chemical menstrua, he employed the thesis structure to outline, categorize, and discuss the basic properties of menstrua. Demonstration experiments served to illustrate claims made directly in the theses. At the beginning of the lectures, he elaborated (in a thesis) on the four types of "causes" that accounted for the action of any menstruum: mechanical, mechanical and repulsive force acting in tandem, mutual attractive forces, and combinations of the three. Only after he had defined the characteristics of each of these classes through theses did he provide examples through demonstration. He offered seven experiments, structured in the demonstrational form, from which he explained how the various causes operated in practice. In the second experiment, he melted "flowers" of sulfur in a closed vessel over a fire and then added drops of mercury, pressed through a linen cloth. He stirred the contents of the vessel with a spatula, and eventually the mass was converted to cinnabar. The point of this demonstration, as he explained, was to show that this menstruum was effected through mechanical action—the motion of fire bringing the sulfur into fusion, the pressing of the mercury through the linen, and the constant stirring—and through a mutual attractive force between the mercury and sulfur.[37]

In terms of the pedagogical aim of the instruments course, Boerhaave's thesis and demonstrational methods were complementary. Both methods produced structurally similar forms of knowledge: pithy statements of fact or generalized conclusions regarding categories of phenomena. Both forms also depended on linguistic manipulation to structure and discipline empirical information. The thesis form required that phenomena be divided into pedagogically efficient bits. The demonstration form depended on language to guide and structure the inferences that the course auditor made from the demonstration. Boerhaave chose to present the various subjects of his lectures through one of the two forms based on the efficacy of each form in addressing the pedagogical challenges posed by the given topic. The primary pedagogical problem for chemical menstrua involved the best method to order and describe the various substances and phenomena that Boerhaave deigned to include within this general heading. Therefore, he chose to rely primarily on the thesis structure, through which he more easily conveyed taxonomic divisions and definitions. Indeed, the first third of the lectures concerned the theoretical principles and classification of menstrua. To this Boerhaave appended sections devoted to discussion of specific types of menstrua common to chemical practice: aqueous, oily, and saline. Each of these sections was further subdivided so that, for example, saline menstrua were divided into simple salts, alkalis, acids, and neutral salts.[38] Conversely, the primary pedagogical problem for fire was the elaboration of vague or generally latent characteristics of a singular, elusive entity. For this problem, he chose experimental demonstrations, because these latent characteristics needed to be generated and clarified by "secondary" instruments (i.e., thermometers) before he could address their significance.

The ultimate goal of Boerhaave's instruments course was to discipline the student into seeing and interacting with the world in a certain manner. The instruments course did this in three ways. First, structured experimental demonstrations guided the gaze of the student toward relevant phenomena and away from others. Through the corollaries, Boerhaave provided an interpretation of the exhibited phenomena and suggested other phenomena that might be explained by the same principles. Second, his methods divided, defined, and classified chemical entities in specific ways, which underwrote his natural philosophical views. For example, Boerhaave's taxonomic classifications of menstrua validated Newtonian theories of repulsion and attraction.[39] Last and most important, the structure of the instruments course provided students with a model for conducting experimental research in chemistry. By displaying demonstration experiments within a series of logical inferences and related interpretations, the order of the demonstrations themselves exhibited Boerhaave's "method of discovery," which he outlined in the introductory lectures of the course. Taken in sum, the instruments course instilled in the student a methodology and philosophical perspective that supported Boerhaave's vision for a chemistry based on experimentally determined principles.

PEDAGOGY AS NATURAL PHILOSOPHY

In putting his instruments course together, Boerhaave played the role of experimental philosopher as well as pedagogue. He invented, routinized, and contextualized experiments as public demonstrations for his students. These experiments had to be disciplined so that they exhibited the proper phenomenon on cue and with enough clarity that Boerhaave could point out the relevant characteristics and draw the necessary inferences. He therefore worked out the demonstrations ahead of time to ensure that they behaved properly when he performed them for his lecture. The process of working out experimental demonstrations was a creative one. The phenomena that he demonstrated were often messy and ambiguous in meaning until he defined them through pedagogical structuring. On occasion, the very process of composing the demonstration lectures sometimes forced Boerhaave to reconsider, clarify, or solidify notions that he had previously held. In creating and ordering his instruments course, Boerhaave was doing experimental chemistry according to his method by finding the precepts and natural principles of the chemical art.

The instrument course took place in the university's chemical laboratory. The first challenge for Boerhaave was to create demonstration lectures that conformed to the physical and social limitations of that space. As indicated in the Leiden course listings (*series lectionum*), he began lecturing at nine in the morning and typically continued for two hours.[40] Students attending the morning's lecture stood in an open area to one side of the laboratory, and Boerhaave delivered his lecture at a table or podium standing among the laboratory furnaces and apparatus. To help him conduct the demonstrations, he had an assistant (*knecht*, literally "manservant") whose salary the university paid. As the official "instructions" to this servant made clear, Boerhaave was to employ him for the preparation and presentation of demonstration lectures and for maintaining the fires in the furnaces, but not for assistance with his own investigations.[41] The demonstrations were performed on a table, with the necessary equipment already set up before the lecture began. The lectures were governed by rules of social conduct and etiquette. The university lecture hall was a space defined by a sharp imbalance of power between professor and student. The professor presented his lessons, and the student was obliged to listen and keep quiet. A student might not agree with the professor's pronouncements, but to vocalize such disagreement during the lecture violated the university statutes.[42]

As stated earlier, Boerhaave's demonstration experiments were not "experiments" in the sense that they tested some hypothesis under question. Rather, they acted as heuristic devices that presented a phenomenon in an unambiguous manner as a means of illustrating that phenomenon to students. As Steven Shapin has pointed out, experimental philosophers made a sharp distinction between "trying" and "demonstrating" an experiment. "Trying" an experiment involved the uncertainty of searching for meaningful results and the labor of routinizing the experiment once such results had been obtained. "Demonstrating" an experiment was the performance of a perfected process in front of an audience.[43] The *experimenta* in Boerhaave's instruments course were demonstrations in this sense, and like any public demonstration, the experiments had to be disciplined to exhibit the proper phenomenon in a controlled manner and on cue. Failure to accomplish this could result in social humiliation and decreased credibility in the eyes of the witnessing audience. The performance of the demonstration concealed the work that went into trying the experiment. First, the material components of the experiment had to be perfected (i.e., the trying proper), and then Boerhaave faced the task of building correspondences between the exhibited experiment and the philosophical or factual claims that he wished to make. This work was primarily literary in that it involved devising arguments that connected the displayed phenomena with concepts that were more general or abstract. Ultimately, the demonstration was integrated into the pedagogical structure of the course, but this process usually posed less of a problem for Boerhaave, because he generally chose experiments that he knew beforehand were relevant to the themes of his course.

Boerhaave contended with various material constraints in designing and perfecting his demonstration experiments. Each experiment had to be concluded fairly quickly because of the limited amount of time for each lecture and the limited attention span of his auditors. Most important, the experiments could not be ambiguous in their results. The audience had to be able to see the important phenomena displayed in the experiment for themselves, and the action of the experiment had to be evident in order to direct their gaze to the proper place. For example, in one extended series of experiments, Boerhaave measured with a large-sized spirit thermometer the heat generated by mixing various liquids. He remarked to his audience that he would have preferred to use a smaller-sized mercury thermometer. He asserted that mercury thermometers were more sensitive and that the large size of his spirit thermometer interfered with the measurement, because the additional thermometric fluid in the instrument absorbed more heat from the fluid being measured in making the reading. He opted for the larger thermometer, because, as he stated, "I wish that you should all fairly perceive the event of the Experiments."[44] Sometimes the laboratory space itself or the attendant conditions of the lecture prevented Boerhaave from pursuing a specific demonstration. In his lectures on fire, Boerhaave addressed the action of burning lenses on bodies. He could not demonstrate such an instrument for his course, however, because the structure of his laboratory did not allow enough light inside the lecture space and with enough regularity for a burning lens to work, especially during the cloudy Dutch winter. Instead, he displayed diagrams of optical instruments, described the geometric path of the rays of light, and calculated the relative intensity of fire at each instrument's focus. The diagrams he showed were of actual instruments described in the *Histoire* of the Académie des sciences: a mirror designed by Villett and a lens fashioned by Tschirnhaus. Boerhaave also portrayed a compound lens instrument designed by Tschirnhaus, and cribbing from a published account, he described the effects of this instrument on a variety of substances.[45]

Boerhaave fashioned his demonstrations out of existing experimental practices. He appropriated most of the demonstrations concerning water (the instrument) and chemical menstrua from well-known chemical operations and procedures. For example, as cited in the last section, he demonstrated the formation of cinnabar from sulfur and mercury as an example of a menstruum affected through both mechanical and attractive forces acting in tandem. Similar demonstrations included the reduction of crude antimony (stibnite) from its ore into *regulus* of antimony (antimony metal) and the separation of copper from silver with lead.[46] Boerhaave was very familiar with these operations, because by 1718 he had demonstrated them in his chemical operations course for fifteen years.[47] Within the context of the instruments course, he used them to illustrate a specific theoretical point. Thus, the main problems for incorporating these operations into the course were ones of presentation and interpretation; he needed to highlight the significant aspects of these operations, provide a theoretical interpretation, and integrate the interpretation within the overall themes and structure of the course. Demonstrations that involved more complex, secondary instruments, such as the burning lens, thermometer, or air pump, posed greater problems for Boerhaave and required more work for redefinition and integration. None of these instruments were construed as traditional chemical instruments in 1718 (although the burning lens was sometimes employed to investigate chemical composition). Boerhaave had to make them so by bringing the conceptual frameworks and experimental practices that governed their uses in line with his vision of chemistry. These instruments had established protocols regarding their operation and the objects they manipulated, and Boerhaave adopted these established practices in constructing his demonstrations. This did not mean that he also accepted extant theoretical claims or experimental programs built around an instrument. Much of his work in trying these instruments involved refining or redefining the theories of their use and developing new experimental lines of inquiry to suit the pedagogical and philosophical needs of his course.[48]

The Fahrenheit thermometer provided a good example of how Boerhaave employed an instrument to guide his theorizing and shape his demonstrations. He did not regard the expansion of fluid due to heat as a central phenomenon relating to fire until he became convinced that the thermometer was an eminently useful instrument for the experimental study of heat. In his earliest chemical lectures he asserted that fire was a subtle fluid, an opinion derived from the Cartesian conception of fire as a form of "first matter" and which he encountered as a student in Leiden in the physics lectures of Wolfred Senguerd.[49] Boerhaave neither discussed thermometers nor mentioned thermal expansion as the "sign of fire" in these first lectures. Only after Daniel Gabriel Fahrenheit (1686–1736) initiated an acquaintance with Boerhaave and introduced him to the principles of thermometry in 1718 did he work to establish the thermometer as a stable chemical instrument. Perhaps what impressed Boerhaave most about these thermometers was Fahrenheit's ability to replicate his scale accurately in new instruments. Most thermometric scales in the early eighteenth century were notoriously unstable, even between thermometers of similar construction. This made the comparison of measurements nearly impossible. Fahrenheit had worked out a method for replicating his scale such that his thermometers regularly read to within a degree of each other.[50] Boerhaave introduced thermometry into his chemical course, because he garnered from Fahrenheit the potential of the instrument to generate the kind of natural philosophical knowledge—empirical, stable, and repeatable—on which he could ground arguments about fire.

Boerhaave did not simply adopt Fahrenheit's program for thermometry but rather reformed Fahrenheit's work to fit into his own pedagogical and methodological goals. Fahrenheit sent four letters to Boerhaave on thermometry during December 1718 and January 1719, exactly when Boerhaave was composing the first section on fire for his instruments course.[51] These letters were filled with descriptions of and data from Fahrenheit's own experimental work on thermometry. At this time he was perfecting his now famous temperature scale, and he revealed to Boerhaave that the novel method by which he replicated his scale in new thermometers was based on careful measurements of the expansion of the thermometric fluid. Fahrenheit represented his scale mathematically in each type of thermometer as the volumetric ratio of the tube to bulb, which was the same fraction that the thermometric fluid expanded over the desired temperature range indicated by the tube. For example, as Boerhaave observed in his lecture notes, the spirit thermometer he used in his demonstration lectures had a bulb volume of 1,933 parts and a tube volume of 96 parts, which indicated that the spirit of wine expanded 96/1,933 of its volume (about 1/20) over 96°F. The same principle applied to the mercury thermometer, and in fact, Boerhaave obtained the correct bulb-to-tube ratio for this thermometer directly from one of Fahrenheit's letters.[52] Fahrenheit's mathematical representation of his scale in terms of fluid expansion directly impressed Boerhaave's thinking on fire. Boerhaave adopted expansion as the "sign of fire" to justify his use of Fahrenheit's thermometer (and Fahrenheit's scale) as the experimental instrument to measure the amount of fire present in bodies. He shaped the theory of the thermometer's action by relating it to his long-time assertion that fire was a subtle, omnipresent fluid. The relative expansion of the thermometric fluid was proportional to the amount of fire it absorbed from the medium surrounding the thermometer's bulb. The Fahrenheit thermometer measured and thus confirmed the existence of fire in bodies, while the theoretical properties of fire explained the action of the thermometer.

Boerhaave also envisioned uses for the Fahrenheit thermometer that differed from the uses its creator intended. Fahrenheit had originally constructed his mercury thermometer as a tool for chemical analysis. He believed that the boiling point, specific gravity, and thermal expansion (over a range of temperatures) of any pure substance were stable and unique. Therefore, if one knew the values of these properties ahead of time, one could potentially measure these properties in an unknown substance and determine its composition. In several of his letters to Boerhaave, he included tables listing the boiling points of various liquids he had tested experimentally with his thermometers.[53] Additionally, he employed his thermometer to fix specific gravity measurements by making such measurements at a set, constant temperature (usually 48°F or 60°F). Fahrenheit explained to Boerhaave how these measurements might be used to determine the composition of an unknown substance. In a letter of December 1718, he described in detail how he determined that a sample of spirit of niter he had purchased was contaminated. He reported that the sample boiled at nine degrees higher than previous samples and had a higher specific gravity. "I conclude[d]," he wrote, "that the same spirits of nitre were not pure, but perhaps adulterated with oil or spirit of vitriol," which was more dense and boiled at a higher temperature.[54]

Boerhaave, however, envisioned the thermometer as a tool for generating philosophical knowledge about the general properties of bodies. He devoted the entire second term of the instruments course (winter 1719–20) to measuring the amount of heat generated (or, in his own terms, the amount of fire attracted or repelled) by the mixing and interaction of chemical species in solution (i.e., menstrua).[55] Although his inspiration for these demonstrations came from Fahrenheit himself, he transformed what Fahrenheit considered to be a phenomenon of passing interest into a research program for natural philosophy and chemistry.[56] For his first demonstration experiment, Boerhaave mixed a quantity of distilled rainwater with an equal quantity of spirit of wine (i.e., ethanol) at the same temperature as the distilled water. The large thermometer placed in the mixing vessel exhibited to his auditors the increase in "degrees of heat" generated by the solution. He performed numerous mixing demonstrations, whose order was determined by the taxonomy of the reagents: vegetable substances first (alcohol, oil of tartar), then animal (salt of urine, sal ammoniac), and finally mineral (mineral acids, metals). The aim of these demonstrations was to produce generalized claims regarding the behavior of fire. For example, Boerhaave performed several variations on the initial distilled water/spirit of wine experiment, altering the concentration of the spirit of wine. Ultimately, he asserted that the purer the alcohol was, the greater amount of heat generated by mixing.[57] These experiments constituted a template for what Boerhaave later called a "complete and certain history of Heat," a reference to the Baconian method that he advocated at the start of his course. He suggested to his students that all "simple bodies" from the three kingdoms of nature should be examined in this way, first by mixing them with other bodies of their "class," and then with bodies of other classes.[58] Boerhaave envisioned the thermometer as the central instrument in an experimental program for examining the relationship between fire (or heat) and chemical interaction.

Boerhaave's use of the air pump in his instruments course similarly illustrated his process of adapting instruments to fit his pedagogical agenda. In contrast to the thermometer, the air pump was an established part of experimental natural philosophy at Leiden in 1721–22, when Boerhaave presented his experimental demonstrations on air. Members of the Leiden arts faculty, such as Burchard de Volder, Wolfred Senguerd, and Willem 'sGravesande, had been presenting air pump demonstrations for over forty years.[59] In the first term of his instruments course devoted to air (1720–21), Boerhaave did not deviate significantly from this established tradition. Building on the pioneering experiments of Robert Boyle and the Leiden natural philosophers, he lectured to his students on the physical properties of air

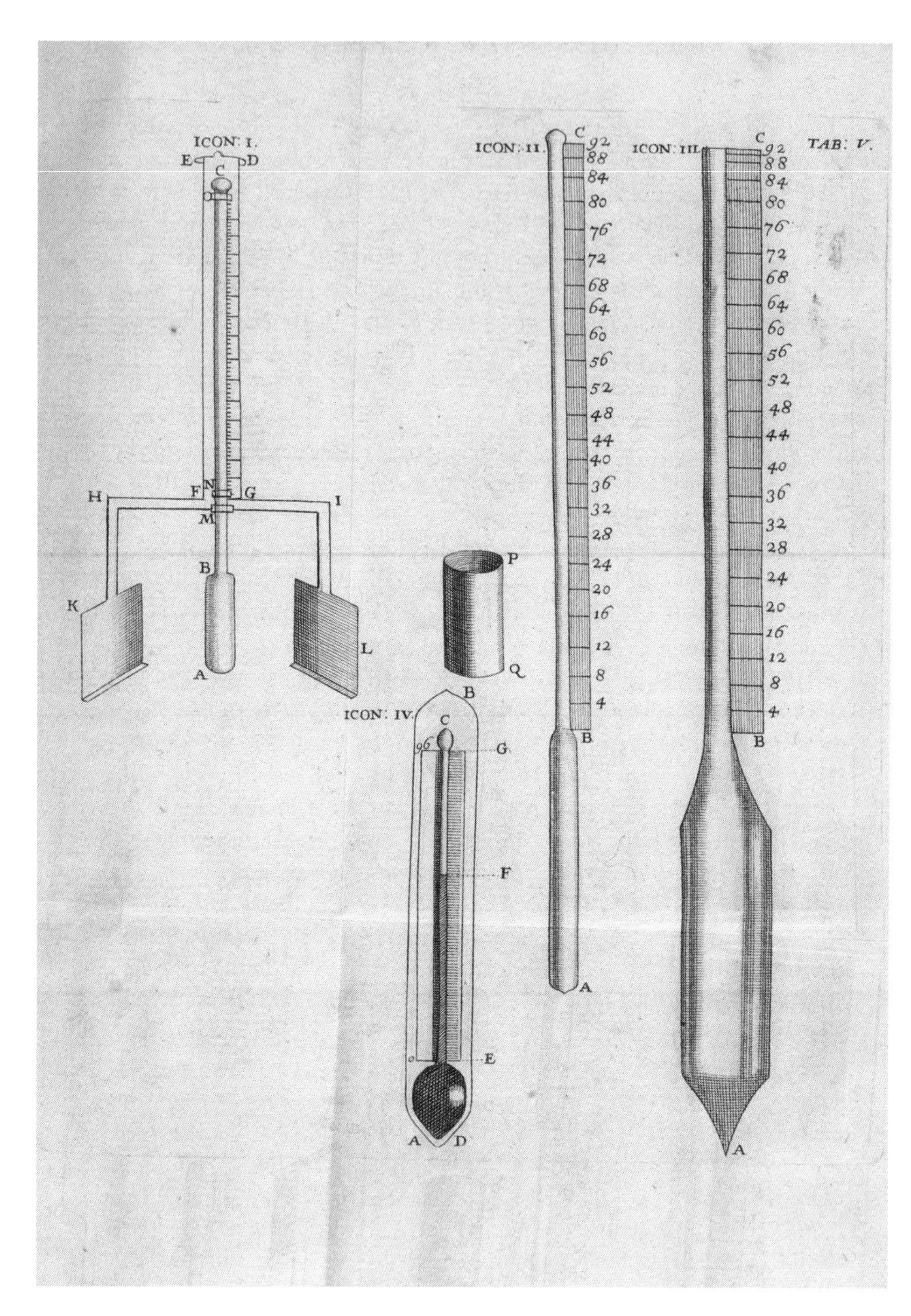

FIGURE 4. Fahrenheit thermometers from the *Elementa Chemiae.* Courtesy of Roy G. Neville Historical Chemical Library, Chemical Heritage Foundation. Photograph by Douglas A. Lockhard.

(weight, elasticity, density) and the vacuum. He cited Boyle's published work on the "spring of the air" several times and conducted an extended discussion on what would later become Boyle's law, which he stated as "the space occupied by the same air is in reciprocal proportion to the compressing weight."[60] Boerhaave, however, did not perform a demonstration of this relationship with an air pump. Instead, he offered a thought experiment: he described a cylindrical tube of air one (Rhineland) foot in diameter and 64 inches high. Boerhaave asserted that the weight pressing on the air inside the cylinder was 2,112 pounds (i.e., the weight of the atmosphere). If one attached to the cylinder a U-shaped vessel filled with 29 inches of mercury, the effective weight pressing on the air doubled, and its volume decreased by half. Boerhaave presented a table of calculations that illustrated how proportional increases in weight reduced the volume of the air to the inverse proportion of the weight.[61] Boerhaave concluded this first term with a discussion of air as a medium for "vapors," such as water, vegetable spirits, salts, and the like.[62]

Boerhaave's second term on air (1721–22) was composed almost entirely of experimental demonstrations. The theme of the second term concerned the extent to which "latent" air could be removed from fluids and solid bodies, and the air pump played a central role in demonstrating this latent air. Boerhaave's twelve demonstration experiments on air were directed toward a theoretical purpose. He designed his demonstrations to show that water could not be transmuted into air. This claim derived from a misinterpretation of Joan Baptista van Helmont's work by Robert Boyle. Van Helmont maintained that water was the first matter from which all other bodies were formed. In his *Ortus Medicinae* (1648), he explained that when water was divided into its smallest parts, it became volatile: more subtle than vapor, steam, or smoke, but more dense than air. He called water in this state a "gas." In Van Helmont's complex theory, a "gas" contained the seminal principle of a specific body that had been reduced to the state of a gas by the power of a "ferment," combustion, or other agent, and thus each gas was unique to the body from which it came.[63] Boyle, however, read Van Helmont as claiming that water could be transmuted into air.[64] Boerhaave, familiar with the texts of Boyle and Van Helmont, designed his experiments to show that any air obtained from water was already present in the water and was not the product of transmutation. His demonstrations attempted to prove this claim by exhibiting the various methods of retrieving dissolved air from water and other liquids. In one set of demonstration experiments with the air pump, he placed a flask of water into the bell of the pump, which he then evacuated. He argued that the bubbles and eventual boiling that was observed were caused by escaping air that had been trapped in the water by the pressure of the atmosphere. He then repeated this experiment with three flasks of water at different temperatures: hot, tepid, and cold. The hot water boiled with the least amount of evacuation, followed by the tepid water, from which Boerhaave concluded that an increased amount of fire (and, thus, mechanical action) helped to dislodge air from the water.[65]

In later demonstrations, Boerhaave modified his air pump to measure the amount of air released from substances as they reacted in vacuo. The inspiration for these experiments also came from Boyle, who performed numerous experiments to generate "steams," "exhalations," and "vapours" from chemical reactions, both inside and outside his air pump.[66] Boerhaave worked out a method to measure the amount of air generated during these reactions. He attached a mercury tube barometer to the pump's glass bell as a means of showing the pressure difference between the inside and outside of the bell, and he rigged the vessels holding his reagents in such a way (which he did not describe) that he could mix them without breaking the seal of the evacuated bell. In the first experiment, Boerhaave mixed 1½ drams of "crabs eyes" (calcium carbonate deposits taken from crayfish) into 1½ ounces of "ardent" vinegar. He reported a violent "ebullition" and that the level of mercury in the barometer (inside the bell) dropped from 28½ "inches" (*pollices*) to 17 "inches" in the first half hour, and after twenty-four hours the mercury had descended to 7 "inches." From this measurement, he concluded that the density of the released air inside the bell was about three-quarters of that of atmospheric air. Knowing the volume of the bell, he calculated that the reaction produced the equivalent of 114 ounces (in volume) of air "in equilibrium with the atmosphere."[67] Boerhaave followed this experiment with four similar demonstration experiments: one in which he mixed chalk and vinegar, one with oil of tartar and vinegar, one with oil of vitriol and oil of tartar, and one with spirit of niter and iron filings. This last experiment, he reported, produced an extremely violent reaction and a copious amount of red fumes (in addition to "latent" air).[68]

In the process of working out these experimental demonstrations, Boerhaave arrived at some novel conclusions regarding the chemical activity of air. In his earliest lectures, he maintained that air was chemically inert, acting only as a mechanical instrument and as a medium for volatile substances.[69] For the first six experiments on air, Boerhaave seemed to support this contention: air could be dissolved in water, but was not concreted in other bodies, and could be released by mechanical means. By the time that he composed his notes on the air generated during chemical reactions, he had changed his mind regarding the chemical role of air. As he recorded in his notes to Experimentum 9, he mixed oil of vitriol and oil of tartar in vacuo, but as he stated in the first two *corollaria,* he had first removed the dissolved air from both liquids by allowing them to sit in the bell of the evacuated air pump for twenty-five hours before he mixed them. The reaction produced a large quantity of elastic air, which might initially have surprised him when he tried this experiment in preparation for this lecture. In keeping with his experiments on the solubility of water, he probably thought that by removing all of the dissolved air first, this reaction would not generate air on the assumption that the usual "effervescence" observed in this reaction was caused by dissolved air forced out of solution by the violent motion of the acid (vitriol) and alkali (tartar) particles. In fact, in the previous *experimentum,* Boerhaave had explained the "effervescence" observed during the mixing of oil of tartar (an alkali) and vinegar (an acid) in this manner.[70] When he composed the *corollaria* to Experimentum 9 in his course notes, at first he merely reported the fact that the experiment produced air, even though he had removed all of the dissolved air. After he had composed the last two demonstrations, he returned to Experimentum 9 and added one additional corollary: "all effervescence continues while friendly bodies rush together in mutual embrace, expell air intimately united with the elements of [those] bodies, and may make . . . air."[71] Thus, trying experiments for his course inadvertently led Boerhaave to the beginnings of a novel discovery: that latent air may be "intimately united" in bodies and may become elastic air again when those bodies are disrupted.

CONCLUSION

Boerhaave's instruments course was a significant milestone in the pedagogy of chemistry because it focused on the experimental demonstration of facts and theories, not techniques of operations. In the process of creating the course, Boerhaave often developed new demonstration experiments and novel theoretical claims to explain the properties and actions of the chemical instruments. Thus, his pedagogical practice was also an enterprise that generated new knowledge and techniques. These demonstrations became paradigmatic examples for the phenomena they displayed and, by extension, for the theories they underwrote. William Cullen (1710–1790), for example, employed Boerhaave's series of thermometric experiments in his chemical lectures at Edinburgh to demonstrate that expansion was the marker of the presence of fire in bodies.[72] Most telling was the fact that Boerhaave's course (and textbook) had the same argumentative force as nonpedagogical texts in experimental natural philosophy. In fact, as I detail in the next chapter, Boerhaave invited the more experienced readers of the *Elementa Chemiae* to read his text as a work of natural philosophy.

The immediate success of his chemical course, however, depended on its pedagogical aspects. Boerhaave attempted to discipline students into becoming experimentalists, which, as he argued publicly in several university orations, also disciplined them into becoming responsible citizens. In reality, few students remained in Leiden over the entire eleven years in which he delivered his instruments course, although one of his protégés, Gerard van Swieten (1700–1772), did.[73] Even though the content of each term's course changed, the pedagogical method and underlying ideology did not. If a student was interested in the content of previous courses, he need only attend Boerhaave's term-length course, which Boerhaave updated as he composed the instruments course (although the term-length course never included demonstration experiments). Ultimately, the instruments course became the "theory" section of the *Elementa Chemiae*.

CHAPTER SIX

Philosophical Chemistry

Near the beginning of the *Elementa Chemiae* (1732), Boerhaave addressed "the service of the [chemical] Art to Natural Philosophy." He remarked on their similarity in approach to nature; both chemistry and natural philosophy studied the sensible properties of natural bodies. Like the natural philosopher, the chemist examined the "phenomena" of bodies and reduced them to a "Natural History." Boerhaave argued that there were two ways of collecting the observations for this natural history: by observing the appearances of bodies standing alone, and by applying different bodies to one another and observing the new phenomena that arose from them. "Chemistry," he maintained, "is almost the only Art that seems suited to cultivate this second and most valuable way of making physical Observations . . . 'Tis this that resolves compound Bodies into their simple parts, and after it has carefully examined them, combines them together again, in order to know what new appearances or powers will thence arise." According to Boerhaave, chemistry, following this methodology, "exhibits to us the instruments by which nature . . . operates" and "pries into her most secret methods of working."[1]

In the *Elementa*, the principles and properties governing the behavior of each instrument were presented in an efficient pedagogical order, along with illuminating examples and demonstrations culled from the mass of facts (i.e., the natural history) from which the principles were deduced. But Boerhaave's account was incomplete. He knew that his chemistry of instruments in its extant form could not explain all phenomena related to chemistry. Further work needed to be done. Once all of the relevant facts were collected and ordered, practitioners would have a full understanding and control of chemical operations. He alluded to this state of affairs at several places in the *Elementa*, but he stated it most directly regarding the chemical menstrua: "If it was possible, now, to reduce Menstruums into an order, according to the difference of their manner of acting, and then be distributed into inferior classes, then the doctrine of Chemistry might be brought to the certainty of a science, and consequently it might always be foretold what would happen in a particular Operation."[2] To assist in achieving this certainty, Boerhaave pointed out to his students the holes and weaknesses of his system of chemistry, and he encouraged them to conduct further investigations designed to establish a complete theory. Chemistry was not a "science" but was on the way to becoming one.

The *Elementa Chemiae* was the first textbook of "philosophical chemistry." In addition to conveying the basic principles and precepts of the chemical art, Boerhaave designed the text to teach the methodology of experiment for chemistry and to promote experimental investigations that would lead to a complete chemical science. This investigative approach to chemistry exemplified what Peter Shaw (1694–1763) called "philosophical chemistry," which would "by means of appropriate Experiments, scientifically explained, lead to the discovery of *Physical Axioms*, and *Rules of Practice*, for producing useful effects" in order to "improve the State of natural knowledge, and the Arts thereon depending."[3] Shaw coined the term "philosophical chemistry" to differentiate the form of chemistry that strove to investigate chemical operations and species systematically (chemistry as natural philosophy) from those forms that sought to apply chemical operations toward some productive end (chemistry as artisanal practice). These two forms of chemistry were interrelated. The aim of chemistry as an art was always to produce some product or other practical result. But following the academic understanding of an art, Boerhaave strove to build the *scientia* of chemistry, that is, chemistry's precepts and principles, which guided practical application. As a textbook of philosophical chemistry, the *Elementa* not only presented the theoretical principles of chemistry but also provided a methodology for how to generate those principles.

The principles and methodology presented in the *Elementa* were the product of twenty-five years of teaching and reforming the teaching of chemistry at the university. The textbook was a composite of Boerhaave's three chemistry courses—his term-length lecture course, his operations course, and his extended instruments course. In building the *Elementa* he appropriated large sections of his courses for the textbook. The short "history" section and long "theory" section, which together comprised volume 1 of the *Elementa*, followed the structure of the first half of Boerhaave's term-length lecture course. His presentation of the chemical instruments, which comprised the bulk of the "theory" section of the *Elementa*, was almost identical in structure and breadth of content to his instruments course. The "practice" section of the *Elementa*, which comprised volume 2, replicated the final version of Boerhaave's operations course, with a few useful theoretical discussions, such as a long discourse on fermentation, added where needed. Just as Boerhaave constantly revised his yearly courses, the *Elementa* constituted the final revision of his notes presented in a comprehensive text with added discussions and commentary not found in his lecture notes. As stated earlier, he maintained that the *Elementa* gave an incomplete account of the chemical art. The book was just one step in a process through which chemistry would eventually be perfected, and Boerhaave provided many signposts to his readers, just as he had to his students, that provided suggestions for furthering the development of the field.

This chapter examines the creation and structure of the *Elementa Chemiae*. First, I examine the circumstances under which Boerhaave composed the *Elementa*. According to his own assertions, he wrote the textbook in response to the publication of a pirated version of his chemical lectures, which fundamentally missed the point of his instrument approach to chemistry. I provide evidence that the *Elementa* was planned to a greater degree than Boerhaave stated. I then discuss how Boerhaave presented in the *Elementa* a methodology for generating chemical knowledge in order to achieve the goals of his philosophical chemistry. In the end, Boerhaave tended to conflate his pedagogical and methodological approaches into one "method." He linked the pedagogical order of the *Elementa* with his method of discovery, arguing that this order represented the best manner of presenting the theoretical principles of chemistry deduced from numerous experimental or observational data. He advocated that his readers create natural and experimental histories in order to establish the theoretical principles needed to complete the "science" of chemistry. Finally, I examine Boerhaave's opinion of the work of three contemporary chemists—Stephen Hales on air, Georg Ernst Stahl on phlogiston, and Étienne-François Geoffroy on chemical affinity tables—as he reported in the *Elementa*. These discussions act as case studies to demonstrate Boerhaave's application of his pedagogical and methodological principles in practice, examine the extent to which pedagogy shaped his thinking, and show that he was, in fact, cognizant of contemporary developments in chemistry.

BIRTH OF A TEXTBOOK

If his own statements are to be accepted, Herman Boerhaave never intended to publish a textbook in chemistry. In the preface of the *Elementa Chemiae*, he lamented that he did not want to bring another chemical book into the world, because there were so many good ones already. He only wished to instruct his students in the "Rudiments" of chemistry and give a few "Examples of the Operations" to those who were interested. He stated that he felt compelled to compose his textbook only after some of his former pupils published a version of his lectures without his consent. In the *Elementa*, he pointedly disowned the book, railing that "The false Notions, Absurdities and Barbarisms, that are imputed to me in every page of that Work, are so abominable, that they will not bear mentioning." He asserted that he appealed to "some Persons in authority," but then "some others" delayed, put off, and opposed his request for action against the book pirates. This situation forced him "immediately" to resign his professorship in chemistry. His friends, he explained, urged him to compose his own textbook to refute publicly the former work and to present to the world the true content of his teachings.[4]

The story elaborated in Boerhaave's preface was shaped by rhetoric to make it appear that he was compelled to write the *Elementa*. Although the facts that he presented were easily substantiated, his account concealed much more calculating motives for publishing his chemical textbook. He actively worked toward the publication of his reformed version of chemistry for several years before he resigned his chair of chemistry. In the five years that separated the appearance of his pirated lectures and his resignation, he completed his instruments course, which became the heart of the theory section of the *Elementa*, and was well into writing the text. In the preface to the *Elementa*, he portrayed himself as a victim in order to discredit the pirated edition of his lectures and deflect potential criticism of his own version. Thus, Boerhaave ultimately used the piracy story as a means to lend credibility to the *Elementa* yet insulate his new pedagogical method from criticism.[5]

The spurious work to which Boerhaave referred in the preface of the *Elementa* was a text titled *Institutiones et Experimenta Chemiae* that identified him as its author. The title page of the original version of this work (1724) bore a Paris imprint, although Boerhaave's student and biographer William Burton alleged that it was actually printed in Leiden.[6] Within three years, this text was reprinted in Amsterdam and Venice, and Peter Shaw and

Ephraim Chambers prepared an English translation, titled *A New Method of Chemistry* (1727), in London.[7] As Boerhaave intimated in the preface of the *Elementa*, the text of the *Institutiones et Experimenta Chemiae* was produced from student notes taken during his chemical courses. As such, the text contained some obvious misinterpretations of Boerhaave's work. For example, at one point in the book the text states that material bodies are composed of five principles: water, spirit, salt, oil, and earth.[8] Boerhaave, however, never advocated this system of principles in his lectures nor in any of his genuine writings. The origin of these kinds of mistakes may never be established, since neither the source of the notes nor the person who collated and edited the text were identified in the book, and no eighteenth-century writer or modern historian has identified them either. By comparing the structure and content of the *Institutiones et Experimenta Chemiae* with that of Boerhaave's various chemical courses, it is evident that the structure and most of the material for the volume were taken from his term-length lecture course (as described in chap. 3) and his operations course. Although by 1724 Boerhaave in his instruments course had presented the expanded discussions and demonstration experiments on fire, air, and water, little of this material appeared in the *Institutiones*.

Despite his frustrations with the *Institutiones et Experimenta Chemiae*, Boerhaave did not expunge his chair of chemistry because of this book. In fact, he resigned his chairs of chemistry and botany to have more time to undertake additional projects, including the publication of the *Elementa*. At the meeting of the Leiden curators and burgemeesters on February 1, 1729, Boerhaave formally requested that he be relieved of his chairs of botany and chemistry. The curators accepted his resignation at the next meeting on February 8, but later they asked him to retain the curatorship of the botanical garden and chemical laboratory until suitable replacements could be appointed.[9] On April 28, 1729, Boerhaave delivered a valedictory oration to the university to commemorate his work in chemistry and botany. In the oration, he provided his audience with several reasons for his resignation, none of which related to the piracy of his chemical lectures. First, he cited reasons of health, stating that his habit of rising early in the morning had exposed him to the risk of cold and brought on a severe attack of gout. Indeed, he had been stricken with two severe illnesses that had him bedridden for several weeks in 1722 and 1727.[10] The main reason for his resignation, however, was that he had grown tired of his teaching duties in chemistry and botany. As he stated, "I realized for the first time that my fear of the enormous labor involved had increased and that, alas, my former enthusiasm for the beginning of a new academic course had diminished."[11] Boerhaave was released from teaching chemistry and botany courses, but rather than rest, he devoted his energies toward other projects. He began an intense period of chemical experimentation on mercury in the university laboratory. He completed a project on a new edition of the collected writings of the Greek medical author Aretaeus of Cappadocia (published in 1731), and he started work on editing a Dutch edition of Jan Swammerdam's *Bybel der Natuur* (published in 1737).[12] He retained the position of professor of medicine, keeping his salary and giving medical courses, including a specialized and extended course on diseases of the nervous system (De Morbis Nervorum, 1730–35).[13] Until the last year of his life, he lectured on the *Institutiones Medicae*, conducted clinical instruction, and offered private medical courses on specific topics.

Although the *Institutiones et Experimenta Chemiae* may have spurred him to write his own textbook, Boerhaave waited for several years to compose the *Elementa*. He concluded his instruments course in April 1728, when he had seemingly completed the history of each of the five instruments in appropriate detail. Additionally, he gave one final lecture and operations course during winter term 1728–29.[14] Given that the entire instruments course had lasted for eleven years—much longer than the vast majority of students stayed in Leiden, one could reasonably conclude that Boerhaave completed the course, because he had already formed the notion of writing a textbook in his mind. He probably began the process of writing the *Elementa* during summer 1728 (at the latest), just after he had finished giving his final lectures on the chemical instruments and was fairly advanced in the writing when he delivered his valedictory oration in April 1729. He completed the manuscript of the text by July 1729 and sent it off the printer, Isaac Severnius. Although the published edition was dated 1732, Boerhaave sent the first copies of the *Elementa* to friends and colleagues in October 1731.[15] This sequence of events indicates that he waited to complete the instruments course, material from which comprised the greater part of volume 1 of *Elementa*, and then taught an additional term of chemistry before he completed the writing of the textbook.

Almost as soon as the *Elementa Chemiae* was published, pirated editions of the work appeared in several European cities. Boerhaave tried to establish the validity of his edition by personally signing every copy printed under his direction.[16] Nevertheless, in 1732 the German publisher J. R. Imhoff printed editions of the *Elementa* in Tübingen and Basel, and a second Leiden edition appeared, printed at Imhoff's expense. Gerrit Lindeboom suggested that

Boerhaave's publisher, Severnius, was in financial trouble and had made a contract with Imhoff (most likely without informing Boerhaave) for the sale of the book in Germany and Switzerland. According to the privilege printed in the Tübingen edition, Imhoff had applied to the elector of Saxony for a license to print the work as early as October 1730.[17] One did not need an insider's edge, however, to publish a version of the book quickly. Editions were also printed in London and Leipzig (first volume only) in 1732, and a Paris edition came out in 1733.[18] Ultimately, about fifty editions, abridgments, or translations of the original *Elementa* were published between 1732 and 1791.[19]

METHOD AND PHILOSOPHY

To refute the "errors" in the *Institutiones et Experimenta Chemiae*, Boerhaave devoted a great deal of effort in the *Elementa* to define his approach to chemistry. He structured the *Elementa* according to the same pedagogical methods he used in his chemistry courses. The aim of his chemistry teaching was to reveal the mechanisms of chemical operations through the study of the chemical instruments (fire, air, water, earth, and chemical menstrua) and their interactions with chemical species. The ultimate goal of the *Elementa* was to present the "principles" of action that would direct chemists in conducting and interpreting chemical operations. Breaking with strictly pedagogical conventions, however, Boerhaave invited the readers of the *Elementa* to view his textbook as work of natural philosophy as well as a pedagogical presentation. He asserted that the principles he presented were generated through the application of "Baconian" natural philosophy, in which individual precepts or demonstration experiments were merely instructive examples of numerous, repeated observations. In areas of chemistry where the guiding precepts were not well established, Boerhaave suggested that his reader should work to construct natural or experimental histories of chemical phenomena from which general principles could be generated. Like the instruments course, the pedagogical *ordo* of the *Elementa* provided models that taught how to construct general principles from empirical and experimental evidence. Boerhaave, in fact, argued that his pedagogical order mirrored the logical constructions of "principles" deduced from numerous chemical facts. Ultimately, the true significance of the *Elementa Chemiae* was that in addition to presenting the principles of Boerhaave's chemistry, the book taught the methodology for generating those principles and suggested avenues for further investigation.

The *Elementa* conflated Boerhaave's pedagogical method (*ordo*) with his methodology for generating new knowledge. Following the humanist pedagogical tradition, the structure of the *Elementa* flowed from Boerhaave's definition of the chemical art. The "theory" section of the text thus began: "Chemistry is the Art, that teaches us how to perform certain physical operations, by which bodies that are discernible by the senses, or that may be rendered so, and that are capable of being contained in vessels, may by suitable instruments be so changed, that particular determin'd effects may thence be produced, and the causes of those effects understood by the effects themselves, to the manifold improvement of various Arts."[20] As with his first chemical *theoria* course, this definition identified the pedagogical loci of the text—objects (i.e., "bodies"), actions (mechanisms of change), uses (how it improved the arts), instruments, and operations—and specified how the topics discussed in these sections logically fit together. Within this pedagogical framework, Boerhaave elaborated the methodological foundation for his chemistry. In a short, "design" section at the start of the *Elementa,* he explained how the pedagogical structure of the text as a whole related to the underlying method of generating the principles of the art. He began with a critique of previous chemists, charging that they tended to "cultivate" their experiments "at random" instead of according to "regular principles." This situation had left chemistry as "a confused heap of events and incidents" in which many crucial particulars were omitted or suppressed. Boerhaave argued that his pedagogical structure imposed order on these disorganized "effects" and deduced "general rules" from them. He maintained that he deduced each of his "theorems" or "principles" from "a multitude of common, incontestable facts, always happening in the same manner." Similarly, he stated that each operation, either as a demonstration in the theory section or process in the practice section, "is considered an example of one of the single cases, from which the general theorem was first deduced."[21] Boerhaave's statements reflected the conflation of pedagogical principles and philosophical proof that was common among eighteenth-century experimentalists. Within this context, a single, well-crafted experiment was sufficient to demonstrate (or prove) a phenomenon, even though that phenomenon might have been discovered only after numerous observations and experiments.[22] Thus, he argued that the inductive framework for discovery, which he advocated for chemistry and medicine in his public orations, was the foundation for establishing the precepts with which he structured his textbook. In effect, Boerhaave posited that these two approaches were two aspects of the same knowledge-generating activity.

The principles with which Boerhaave constructed the "theory" section of the *Elementa* derived from his understanding of the chemical instruments—fire, air, water, earth, and chemical menstrua—and how they interacted with chemical species. Following his earlier *theoria* course, he justified this approach to chemistry by arguing that the traditional systems of chemical principles were invalid. His position stemmed from a common seventeenth-century critique of chemical analysis by fire. Critics of analysis by fire (notably Robert Boyle) argued that chemical operations often did not resolve bodies into their elements or principles, as many chemists claimed, but created new compounds.[23] In the new "actions" section in the *Elementa*, Boerhaave greatly expanded his arguments against chemical principles from his earlier lecture course by explaining chemical action at the particulate level. He argued that all chemical change was the result of the motion and rearrangement of particles caused by mechanical disruption and the attraction of various species of particles for one another.[24] He warned that one must not assume that chemical analysis separated a body into the elemental particles that existed in that body before the analysis: "For since the same powers that disunite these corpuscles [usually fire], may produce in them likewise a very great alteration, we shall often fall into error if we suppose that the compound bodies in reality do contain these very Elements. And indeed, upon the resolution of Bodies, there often arises in the parts of them, new virtues, which were never discovered themselves by any effect in the bodies while they were entire; of which there is an infinite number of examples." Indeed, Boerhaave asserted that no chemist had been able to isolate pure water or air or earth and that most of the extracts from plants and animals were capable of further division, indicating that they were not elemental. Even when one had obtained these extracts, it was exceedingly difficult to recompose these elements back into their original substances, as he pointed out by citing the analysis of blood and wine. Thus, he concluded, "it is necessary to fix some sure limits to our Art," and he suggested that chemists should confine themselves to examining the "determined effects" of individual operations.[25]

Despite this seeming rejection of the chemical "elements," Boerhaave found that he could not expunge all the hypothetical principles that shaped earlier explanations of phenomena central to chemical practice. The "theory" section of the *Elementa* was full of entities that he referred to as "principles" or "elements." In fact, he presented a list of "chemists' elements" immediately following his general critique of those same elements. Included on the list were four of the instruments—fire, air, water, and earth—and three others: "Alcohol of wine, Mercury (of metals), and the *Spiritus Rector* of every body."[26] Boerhaave then elaborated on what he meant by "element." He argued that although they "do appear exceedingly subtle and durable," none of these entities could be obtained in pure state by chemical methods. Thus, one could not posit with certainty that, according to the traditional definition of an element, any of these entities were constituents of bodies. For the four instruments on the list, this point was primarily academic, since in Boerhaave's chemistry they functioned as tools and in the main did not play an important role as constituents of bodies.[27] For the other three elements, however, the problem of analysis led to a larger, theoretical problem. According to Boerhaave's definition, chemistry "manipulates bodies that are discernible by the senses, or that may be rendered so, and that are capable of being contained in vessels."[28] The last three elements did not follow this rule; they were subtle or elusive substances. In Boerhaave's chemistry and, indeed, in the traditional chemistry from which they derived, these "elements" conferred to substances in which they were compounded specific properties that explained important phenomena. For example, "alcohol of wine" was the purest form of Boerhaave's *pabulum ignis*—his principle of inflammability—that was obtainable through chemical means. Derived from the traditional, "sulfur" principle, the *pabulum ignis* functioned as the material cause of combustion. Bodies containing *pabulum ignis* would burn if given the proper application of fire; those without it would not.[29]

Boerhaave conducted a series of experimental investigations designed to isolate and examine each of these subtle elements in an attempt to place them on a solid theoretical foundation in accordance with his own methodological rules. As described in the *Elementa*, he began his investigation of the *pabulum ignis* with a consideration of the products one obtained through the dry distillation and calcination of vegetable matter, testing which products of this analysis were combustible. After reviewing each product—water, spirits, oils, smoke, salts, and other items—he concluded that the combustible products were the oily ones.[30] Not all oils, however, burned easily. Oil of turpentine, for example, was difficult to burn, unless it had been "attenuated" though art: successive distillations, putrefaction, or fermentation. For Boerhaave, the most attenuated oil, and thus the purest form of the *pabulum ignis*, was alcohol of wine, which was the product of both fermentation and distillation.[31] In the hope of observing or collecting the *pabulum ignis* itself, he conducted a series of experiments in which he burned the purest alcohol that he could make under a bell jar designed to collect the products of combustion. Despite his efforts, his experiments produced only water with no other detectable product. After a protracted discussion in the text, Boerhaave concluded that the *pabulum ignis* must have been fixed in the water (thus, alcohol is water combined with the *pabulum*) and released by the action of the fire. The *pabulum* itself must be "so vastly fine, that it is dissipated into the chaos of the Atmosphere, and goes beyond the reach of our senses."[32] He undertook similar investigations for his *spiritus rector*, which was a subtle principle that gave the essential oils of plants their specific characteristics, and his "Mercury," which was the root of metals. Both of these endeavors yielded similar, inconclusive results.[33]

As his failure to isolate the *pabulum ignis* and other chemical principles suggests, Boerhaave was very much aware that his chemical system was incomplete. In effect, chemistry was an unfinished art, meaning that many of the principles and entities of chemistry were yet to be discovered or understood completely. As a means of collecting the experimental facts needed to complete the art, he encouraged his readers to continue the project of finishing the experimental history of chemistry. Throughout the *Elementa Chemiae*, Boerhaave maintained the mentoring tone of his chemical lectures, posing questions, revealing difficulties, and suggesting avenues for further research. One example of this mentoring was found in the section on fire at the conclusion of his presentation of a series of demonstrations in which he measured the heat evolved though the mixing of different chemical substances. He suggested that those who had the inclination ought to obtain a mercury thermometer from Fahrenheit and pursue their own experiments, first by mixing "simples" of the same class and then moving on to "simples" from different classes. By carefully noting the phenomena that arose, one would generate a "compleat and certain history of heat." In another part of the text, Boerhaave discussed the life-giving qualities of air, suggesting that common air contained a "certain hidden virtue," which the alchemist, Michael Sendivogius, called the "Food of life." Boerhaave asserted that no physician or philosopher had yet discovered what this food was or how it acted. In the end, he tantalizingly suggested, "Happy [will be] the person who shall happen to discover it."[34]

Nevertheless, the *Elementa* could be seen as the culmination of Boerhaave's own efforts at chemical reform. He elaborated arguments and examples from materials and experiences that he had collected over the years from a variety of sources. Much of this material probably derived from oral explanations of the theses that Boerhaave presented to his students but did not record in his notes. His lecture notes only provided the skeleton and direction for the actual lecture. He spent much of the lecture in exposition on a given topic, citing examples and arguments at least partially from memory. One advantage to leaving his lectures in this semi-fluid format was that (in the case of his standard lecture course) he could address new developments in the chemical arts, or changes in his own opinions on topics, each time he presented the course, without completely restructuring his notes. Indeed, Boerhaave's notes for his term-length course were replete with comments in the margins and added theses reflecting new topics or examples that he incorporated into his lectures after he had prepared the original notes in 1702.[35] Major changes, however, required restructuring. In 1718, when he became interested in thermometry and built his instruments course around it, he also rewrote the first half of the section on fire in his term-length course to reflect his new approach and summarize his new claims.[36] In all probability, Boerhaave also changed the oral examples that he presented during his lectures, but unfortunately no good record exists to show how these oral examples developed over twenty-six years of lecturing.

Translating the performance of experimental demonstrations from his instruments course into a textual form presented Boerhaave with a more complex problem. An essential component of his demonstration experiments was visual: students were conditioned by the performance and accompanying lecture to see a phenomenon in a certain way. Because this visual component was missing in the medium of the book, Boerhaave resorted to what Steven Shapin has described as "virtual witnessing" by a detailed description of the experiment.[37] To make up for the lackproviding of visual information, he often employed illustrations or geometric diagrams to help convey the visual and spatial sense of his demonstrations. Of course, such descriptions, especially those with accompanying geometric diagrams, presented the relevant phenomenon in a much clearer light then when one experienced an actual demonstration, because these textual forms eliminated the visual noise that was present in the personal observation of the experiment. Nevertheless, in the *Elementa*, Boerhaave presented his demonstrations in the same demonstrational form with which he structured his lecture notes. Each *experimentum* was numbered according to the same order as it appeared in his notes. He followed the detailed description of the experiment with a numbered set of *corollaria*, in most cases identical to those he originally listed in his notes. These *corollaria* conveyed the proper inferences that the reader was to derive from the described experiment.

Within this formal structure, Boerhaave elaborated on important points with additional examples and descriptions. These elaborations usually took the form of addenda to the *corollaria*, but occasionally he added additional examples to the experiments as well. To see how this worked in practice, let us examine Experimentum 1 in the section on fire of the *Elementa*, which described the expanding bar demonstration (also Experimentum 1) in the fire section of the instruments course. Boerhaave started with a thesis: "Fire expands the hardest bodies in all their dimensions so long as it is contained in them." He then recounted the same experiment he had performed for the instruments course. He described two iron bars of equal dimensions, heated one in a furnace, then compared the two bars, explaining how the heated bar was visibly longer and its circumference measurably larger than those of the cool bar. This was the end of the experiment in the instruments course, but to the account in the *Elementa* he added a description of how one might measure the expansion of the bar. Boerhaave's description included a diagram (in a plate at the back of the book) of a type of brass caliper and an explanation of how to make measurements with the device.[38] He then presented seven *corollaria*, which replicated the seven *corollaria* that for this experiment in his notes from his instruments course.[39] He added to the first corollary, which stated that all solid bodies expanded when heated, a caution to the reader not to assume that all bodies expanded at the same rate or in the same proportion. To illustrate this point, he related a story about how he had commissioned the instrument maker Daniel Fahrenheit to construct two thermometers, one spirit and one mercury, which read to the same degree under the same conditions. When he received the thermometers, they did not read in agreement. He informed Fahrenheit of the problem, who immediately began to investigate the error. Eventually, Fahrenheit reported that he had fashioned the two thermometers out of two different kinds of glass ("Bohemian" and "Dutch"), and he had discovered that one type of glass did not expand as greatly as the other when heated, thus accounting for the differing readings.[40] Examples and additional descriptions such as these two were not essential to the pedagogical purpose of the textbook, but they served to give the *Elementa* an added flavor that probably resembled Boerhaave's actual lectures.

Boerhaave also made a concerted effort to integrate new material into the *Elementa* that he never discussed in his chemical courses. For example, when Boerhaave was in the midst of composing the *Elementa*, he wrote to Fahrenheit asking the instrument maker to supply him with additional information on his thermometers and inquiring into some of his experimental pursuits. Fahrenheit, eager to have his work publicized in Boerhaave's textbook, gladly supplied the details to the Leiden professor. Boerhaave extracted from Fahrenheit's correspondence the example just described regarding the two thermometers constructed with different glass. Other material that he incorporated into the *Elementa* included some of Fahrenheit's experiments on the "artificial production of cold" and on mixing liquids at different temperatures to obtain a mean temperature. In both cases, Boerhaave discussed Fahrenheit's work in the *corollaria* of extant demonstrations. In the former case, Fahrenheit described how he added successive quantities of aqua fortis to ice water and was able to achieve temperatures of −40° on his scale. Boerhaave reported Fahrenheit's experiments in the section on "fire" as part of a discussion of freezing and the limits of cold found in Experiment 4, in which he described the contraction of air in an air thermometer on cooling. The fourth *corollaria* for this experiment had originally contained an exposition of the "greatest natural cold" (*frigus naturale maximum*), and Boerhaave added Fahrenheit's work as an example of the greatest "artificial" cold.[41]

In a few instances, Boerhaave added new sections to the *Elementa* that were not present in his lectures in order to accommodate topics that he felt were important but that he could not easily integrate into his extant pedagogical structure. One notable example was a section on phosphorous that he added to the end of his discussion of fire. He added this section in recognition that phosphorous had become an increasingly important subject in natural philosophy and chemistry since the later seventeenth century.[42] Following his academic method, he defined "phosphorous" as a body that "takes fire and flame" from contact with the air alone, and he included two different recipes for making such a substance—one attributed to the German alchemist Johann Crafft and one to Wilhelm Homberg. The spontaneous flaming of phosphorous, however, was an anomaly in Boerhaave's theory of fire. Although he attempted to explain this phenomenon by invoking principles he had previously established, ultimately Boerhaave classified phosphorous as its own category, exclaiming, "Who can pretend to set bounds to the power of fire?"[43]

ADDITIONS AND REFUTATIONS

In composing the *Elementa Chemiae*, Boerhaave was well aware of contemporary concerns and researches among his fellow chemists and natural philosophers. His response and interpretation of their work and claims constitute many of the additions and examples that appeared in his textbook. Because Boerhaave structured the *Elementa* to reflect his own view of what chemistry should be, namely, a collection of natural and experimental histories from which one extracted the natural principles and mechanisms of chemical operations, he integrated this material into the text where he deemed it appropriate for his system. He divided other chemists' work into manageable claims, distributed the claims among the relevant pedagogical loci, and applauded or refuted the claims as they supported or contradicted his own. As a result, phenomena that other chemists grouped together or related to one another through a theoretical construct (such as a system of chemical principles or alternate method of ordering) might be separated by Boerhaave's particular method and made to serve as examples within his instrument approach to chemistry.

As a result of Boerhaave's approach, the work of other chemists that historians have identified as central to the development of chemistry in the eighteenth century often seemed lost or overlooked in the *Elementa*. In part this view stems from the fact that much of the history of chemistry in the eighteenth century has been written from the perspective of developments later in the century—the chemical revolution—looking back.[44] The *Elementa* provides an excellent opportunity to examine the status of several important debates in chemistry from the perspective of the changing field early in the eighteenth century. The rest of this chapter examines Boerhaave's views as expressed in the *Elementa* on three areas of concern that historians have come to regard as hallmarks of eighteenth-century chemistry: the pneumatic chemistry of Stephen Hales (1677–1761), the phlogiston theory of Georg Ernst Stahl (1659–1734), and the notion of chemical "affinity" as expressed through the table of *rapports* of Étienne-François Geoffroy (1672–1731). By examining these areas through the lens of the *Elementa*, I show that Boerhaave was aware of contemporary work in chemistry and engaged in experimental and interpretive work on each of these topics. This examination reveals that many of these "hallmarks" of eighteenth-century chemistry simply had not been established as such by the late 1720s. The elevation of these topics was the work of later chemists (and historians) who adopted the earlier work to support their own disciplinary agendas.

Hales on the Reactivity of Air

Boerhaave's opinion on the chemical reactivity of air has always been a matter for contention among historians. The predominant view has been that Boerhaave did not accept that air could be fixed in bodies until he read Stephen Hales's report on experiments to extract air from solid substances, found in his *Vegetable Staticks* (1727). This interpretation of the Leiden professor's change of heart originated in the eighteenth century, asserted by no less an authority than Lavoisier himself.[45] I have shown (in chap. 5) that by March 1722, five years before the appearance of Hales's book, Boerhaave became convinced that air combined "in intimate union" with bodies. Although this evidence establishes that Boerhaave came to this conclusion independent of Hales, one may also ask why he did not announce this "discovery" at the time he made it. An examination of his discussion of the reactivity of air in the *Elementa* demonstrates that Boerhaave did not see his work as a great discovery. In effect, his concerns were not the same as those of later pneumatic chemists, who credited Hales as the founder of their field.

When Boerhaave composed his demonstration lectures on air for his instruments course, and later, when he wrote the *Elementa*, he was immersed in the debates on air that had come into print over the previous century. In Boerhaave's mind, the work of Joan Baptista van Helmont and Robert Boyle dominated these debates, and he constructed his experimental examination of air in response to the work of these two authors. As I related in chapter 5, Boerhaave designed his demonstration experiments to show that water (or other bodies) could not be transmuted into air, as Robert Boyle had argued against the claim of Van Helmont. Boerhaave's demonstrations on the solubility of air in water were supposed to show that the air generated by some chemical operations was not transmuted water but rather was dissolved air forced out of solution by mechanical action. Only when Boerhaave found that oil of vitriol mixed with oil of tartar generated a large amount of air, despite having removed the dissolved air from the samples in vacuo, did he conclude that some air must be "united" with the reacting substances.[46] In the *Elementa*, he clearly placed his examination of this air within the Helmont/Boyle debate. After describing the oil of vitriol and oil of tartar experiment, he queried whether the generated air was what Van Helmont termed a "gas sylvestre," a body resolved into its "minutest parts" and "by a real transmutation, [changed] into this elastic Air." Conversely, Boerhaave asked whether this air was "common elastic air" or "something else very much resembling it, and yet not perfectly the same."[47]

After Boerhaave had concluded his descriptions of the experiments on air, he added a discussion in which he established his conclusions on the air produced by chemical reactions. He began by summarizing from the work of Van Helmont and Boyle on all of the types of reactions—fermentation, combustion, putrefaction, distillation—that generated air. From these examples, Boerhaave concluded that "this elastic Air concurs [*sic*], as a pretty considerable and remarkable constituent part in the composition of almost all kinds of bodies." He then explained that this "elastic Air" did not "produce the effects of air" while united with bodies, but upon resolution through the action of fire, the air regained its former elasticity, until it was united again with "non-aerial" particles and concreted into one mass. Ultimately, he directed the implications of this theory toward his original aim, which was to argue against the transmutation of non-aerial substances into air. He thus concluded, "In all these cases, therefore, air is immutable." By this statement, Boerhaave also implied that there was only one kind of air. Although he never stated this claim overtly, he also never referred to any type of air other than "elastic air," except in the query cited earlier. In a summary paragraph at the end of the discourse on air, however, Boerhaave suggested that the volatile particles that were present in free air might also be concreted in bodies with the air itself and cause "an infinite number of effects."[48] This statement was probably an attempt to account for the various types of fumes or vapors that he, along with Van Helmont and Boyle, had observed in experiment.[49]

Boerhaave did not claim to have discovered that air concreted in bodies (either in the *Elementa* or earlier) primarily because he did not believe that he had discovered anything novel. From the reading of the *Elementa* presented here, one can conclude that he interpreted his own work as refining and extending that of Van Helmont and Boyle. Boerhaave ultimately did not agree with either of their theories, which attempted to account for the generation elastic air through the "effervescence" of certain chemical species. He clearly credited the two for identifying and defining an important category of phenomenon, and for Boerhaave, this was the important work. He saw his own role as evaluating the mass of facts that Van Helmont and Boyle had collected and then placing the facts into a logical order from which he could draw conclusions. The aim of Boerhaave's pedagogical order with which he presented this material followed the agenda set by Boyle: to demonstrate that air was conserved and immutable. Boerhaave invoked the work of Stephen Hales within this context. He argued that the immutability of air was proven through the "resolution and composition" of bodies that contained concreted air, and that pedagogical completeness demanded that he give examples of both. He stated that he had "formerly made a great many Experiments to this purpose: But having seen, and to my advantage perused, a very elaborate treatise, published about two years ago by the famous Dr. *Stephen Hales*, called *Vegetable Staticks*, in the sixth chapter of which he has . . . given an elegant account of his Experiments, which illustrate this affair, and in short quite complete it, I choose rather to refer you to that work."[50] Thus, by his own account, he recommended Hales's work on air to supplement his own, because it was the most complete account available. At no time, however, did he ascribe a discovery either to Hales or to himself.

Boerhaave's work on the concretion of air in bodies was later overshadowed by that of Hales for several reasons. As Boerhaave himself implied in the *Elementa*, Hales's treatment of the subject was much more comprehensive than his own. When pneumatic chemistry became important in Britain during the mid-eighteenth century, Hales's *Vegetable Staticks* provided a wealth of experimental and theoretical resources in the local vernacular. The impetus, however, for the British studies in "aeroform fluids" was not Hales's work directly; rather it derived from the burgeoning interest in the analysis of spa waters, especially Peter Shaw's work during the 1730s on the "spirits" contained in these waters. Although Hales contributed to these investigations, Boerhaave had nothing to say on them.[51] The analysis of spa waters also seemed to spark the French interest in Hales's work, although his main French promoter, G. F. Rouelle (1703–1770), who incorporated Hales's work into his extremely popular lectures at the Jardin du roi, did not come to Hales through this route.[52] Boerhaave, along with his colleague on the Leiden philosophy faculty, Pieter van Musschenbroek (1692–1761), played an instrumental role in spreading the popularity of Hales's work on the Continent through its favorable mention in the *Elementa*.[53] Antoine Lavoisier (1743–1794), however, was the researcher who firmly established Hales as the "father of pneumatic chemistry." He took Hales's work and portrayed it as a predecessor of his own, primarily because the French chemist could cite several of Hales's experiments as confirmations of his own theories.[54] By contrast, Boerhaave's experiments against the transmutation of air did not speak to Lavoisier's theoretical interests. His ignorance regarding the timing of Boerhaave's lectures on "air," coupled with some seeming inconsistencies on the subject in the *Elementa*, lead Lavoisier to conclude that Boerhaave based his understanding of air on Hales's work.

Stahl on Phlogiston

As Gerrit Lindeboom has observed, Boerhaave did not mention Georg Ernst Stahl's "phlogiston" anywhere in the *Elementa Chemiae*.[55] Indeed, most historians of chemistry have counterpoised Stahl's theory of phlogiston with Boerhaave's notion of a *pabulum ignis*, portraying the two ideas as theoretically diverging approaches to similar problems. Boerhaave's *pabulum* was the material cause of inflammability, but it needed to interact with instrumental fire, a separate entity, for combustion to take place. By contrast, Stahl's phlogiston, at least as interpreted by his followers, was elemental fire "fixed" in the inflammable body, although Stahl himself resisted this interpretation.[56] Stahl's phlogiston was also involved in a larger variety of phenomena. In the classic example, he maintained that the calcination of metals was a chemical process in which phlogiston was driven out of the metal, leaving a metallic earth or calx. Boerhaave, by contrast, saw calcination as the disruption of a metal's particulate structure by fire, implying that the calx was simply the attenuated body of the metal.[57] The differences between the two entities stem from the different methodological approaches to chemistry of Stahl and Boerhaave. Stahl's unification of a large number of phenomena (calcination, combustion, color, etc.) through the phlogiston theory was antithetical to Boerhaave's "natural history" approach to structuring the chemical art. Yet, despite these differences, Stahl's phlogiston and Boerhaave's *pabulum* derived from the same interpretive tradition that focused on the role of the chemical principle "sulfur" around the turn of the eighteenth century.

Chemistry was not the main arena in which Boerhaave and Stahl had divergent opinions. In his medical courses at Halle, Stahl promoted a form of vitalism in medicine. He taught that the human body was more than the sum of its parts; it was a holistic organism in which its material parts and the mind (or soul) were united. Boerhaave embraced a mechanical understanding of the body (as described in chap. 4) and upheld the orthodox, Calvinist view that the mind and body were distinct entities. In fact, Stahl's Pietist faith and its implications for his medicine may have rankled Boerhaave as much as any of Stahl's specific medical or philosophical claims.[58] Nevertheless, Boerhaave rarely criticized Stahl in public or print, preferring as was his policy to avoid scholarly polemics. Instead, he gave ardent support to Friedrich Hoffman (1660–1742), Stahl's onetime rival on the medical faculty at Halle. Hoffman advocated a version of mechanical medicine in which Boerhaave saw much of his own approach.[59]

Perhaps because of his dislike of Stahl's medical philosophy, Boerhaave was acquainted only vaguely with Stahl's chemical work. In the *Elementa*, he often failed to differentiate Stahl's claims from those of Johann Joachim Becher (1635–1682), who was one of Stahl's early influences. For example, in the *Elementa*'s section on "water," Boerhaave attributed to Stahl the claim that the particles of water may be attenuated by fire to such a degree that they may pass through the walls of glass vessels. In the next sentence, he related this claim to a similar view by Becher, who asserted (as did Stahl) that water may be given a "corrosive faculty" through successive distillations.[60] Such associations, as well as the fact the Boerhaave never mentioned phlogiston in the *Elementa*, probably derived from Boerhaave's ignorance of much of Stahl's chemical work. The only work he cited in the *Elementa* was Stahl's *Fundamenta Chymiae* (1723). This chemical textbook was based on Stahl's lectures from the 1680s, before he had devised his phlogiston theory. In this book (and his early lectures), Stahl openly acknowledged his intellectual debt to Becher and adopted Becher's system of four chemical principles: water and three types of earth.[61] It is unclear whether Boerhaave was familiar with any of Stahl's other published work, although this alone does not provide evidence that he was unfamiliar with the concept of phlogiston. Other sources known to Boerhaave discussed phlogiston during the 1720s, including Daniel Fahrenheit in his public lectures given in Amsterdam and Étienne-François Geoffroy in a paper published in the *Mémoires* of the French Academy. Neither of these sources, however, identified Stahl as a radical innovator (Fahrenheit did not even record his name in his lecture notes) but rather placed phlogiston within existing chemical traditions concerning the chemical principle "sulfur."[62] These examples suggest that Boerhaave, like many of his contemporaries, tended to incorporate Stahl's work and views into existing chemical traditions.[63]

This interpretation is justified by the fact that Stahl devised the theory of phlogiston within the established research tradition of examining the composition of mineral sulfur. The nature of sulfur, sulfurous bodies, and their relationship to traditional chemical and alchemical notions of the sulfur principle constituted an important area of research for many chemists around the turn of the eighteenth century. A few decades before Stahl undertook his work, Robert Boyle and Johann Rudolph Glauber (1604–1670) both published accounts of methods to synthesize common sulfur by combining acid salts and sulfurous substances (such as oils or charcoal).[64] In 1703, Wilhelm Homberg published an "analysis" of common sulfur, arguing that it was composed of an acid salt, an earth, and the sulfur principle, and in the very next year, Étienne-François Geoffroy published a *mémoire* describing a "recomposition" of sulfur from Homberg's ingredients.[65]

Stahl composed the first work in which he discussed phlogiston, *Zymotechnia Fundamentalis* (1697), within this tradition of examining the chemistry of sulfur. Although the *Zymotechnia* was a book on fermentation, he also announced a "new experiment" on the synthesis of "true sulfur" (*sulphur verum*) achieved by heating vitriolated tartar with charcoal. The significance of this operation lay in Stahl's interpretation: he argued that his sulfur was composed of vitriolic acid (from the vitriolated tartar) and "phlogiston" (from the charcoal).[66] As evident from this first example, Stahl's phlogiston held the same place in the composition of common sulfur as did the sulfur principle and performed the same role—the principle of inflammability. In subsequent works, he associated his phlogiston with Becher's inflammable earth (*terra pinguis*) and thus redefined the calcination of metals and the combustion of sulfurous bodies as related phenomena.[67]

Boerhaave's *pabulum ignis* derived from this same tradition. He had adopted the term *pabulum ignis* from Johannes Bohn in his earliest chemical lectures to denote the subtle, sulfurous agent that acted as the material cause of inflammability.[68] In his instruments course, Boerhaave described a meticulous experiment in which he burned a piece of wood and then, using the products of combustion, attempted to isolate the *pabulum ignis* in a pure state. As described earlier in this chapter, he concluded that oil from the wood contained the *pabulum*, the purest example of which was the highly rectified (i.e., distilled with water) essential oil of the plant. The other products of combustion that appeared combustible, such as charcoal or soot, were only so because they contained trace amounts of oil.[69] Boerhaave suggested that a more effective method to isolate the *pabulum* was through fermentation. Just as distillation removed the dross matter from the essential oil, Boerhaave argued that fermentation "attenuated" the oils of a plant, producing the *pabulum* as alcohol. He asserted that distilled alcohol was the purest from of the *pabulum*.[70]

Since their theories derived from the same tradition, Boerhaave and Stahl largely agreed on the composition of inflammable substances. In his instruments course, Boerhaave performed a series of demonstration experiments on the combustion of alcohol. In one experiment, he set flame to distilled alcohol under a bell jar and collected the product of this combustion, which he identified as water. He observed that this combustion did not produce any soot, smoke, or feces, which indicated to him that there were no impurities in the alcohol. From this experiment, Boerhaave concluded that the combustion of alcohol constituted the separation of water from the *pabulum ignis*, which "becomes so vastly fine" that it is "beyond the reach of our senses."[71] This conclusion was significant because it was the same conclusion that Stahl had reached two decades earlier. In his *Zymotechnia Fundamentalis*, Stahl argued that the "oily" spirits produced during fermentation were composed of water in weak union with his subtle phlogiston.[72] Boerhaave also agreed with Stahl regarding the composition of common sulfur. In the *Elementa* he attested that sulfur was composed of *pabulum ignis* united with spirit of vitriol, as indicated by the fact that sulfur burned under a bell jar produced oil of vitriol. In fact, Boerhaave viewed these two operations as analogous, as he expressed in a *scholium* following his combustion experiments: "If these things are really true . . . then Alcohol would very much resemble Sulfur, as both of them would be totally consumed in the Fire; both of them would yield a blue flame; be resolved into one invisible part, that is inflammable; and whilst burning, yield another part that extinguishes Flame; which in Alcohol is inactive water; in Sulfur, an exceeding acid Salt of Vitriol."[73] Indeed, Boerhaave's *pabulum* and Stahl's phlogiston were interchangeable entities in certain contexts.

One set of differences between Stahl and Boerhaave arose from their differing views regarding the nature of the metals. As has been well documented, Stahl maintained that metals were compounds of phlogiston and an earth, although in some of his writings he seemed to suggest that a "mercurial spirit" was also required.[74] During the calcination of a metal, phlogiston was released into the air, leaving "ash" (*cinis*, i.e., calx), which was primarily composed of earth. Stahl, however, seemed to be only one of several French and German chemical practitioners who, building on the work of Becher, held that metals were composed of "earth" and some form of inflammable principle.[75] Boerhaave was one of the few prominent chemists at the time who contended that most metals contained no earth. His position on this issue derived from his interest in alchemy, as his theory of metals was a modified version of the Geberian mercury/sulfur dyad.[76] In the section on earth in the *Elementa,* Boerhaave explained that "the most ancient chemists" posited that metals consisted of an "exceedingly homogeneous mercury" and a subtle, metallic sulfur. The perfect metals, gold and silver, contained only these two principles, but the baser metals also contained another material that was "sub-pinguious," somewhat inflammable, and could not resist fire. Boerhaave observed that "moderns," on the basis of their experimental evidence, have posited that an "earth that will vitrify" also entered into the composition of metals, an obvious reference to the followers of Becher.[77] Boerhaave suggested that these chemists have misinterpreted their results. As an example of such misinterpretation, he cited an experiment by Homberg in which the French chemist claimed to generate an earth from common mercury by shaking it for an extended period of time. Boerhaave reported that he too tried this experiment, but the black powder that he generated was not an earth but a "wonderful metalline production" that he could convert back into pure mercury through the action of fire.

He also revealed that he attempted this experiment with alloys of mercury and various metals, obtaining the same results.[78] For Boerhaave, there was only one type of earth, and as an instrument its function was to "fix" volatile substances, such as acid and alkaline spirits. The calx of a metal was not earth but a different form of the metal that had been altered by fire. As evidence for this claim, he pointed to the fact that metallic calces could be "converted" into a glass, which, he argued, "everyone who is acquainted with these things knows very well can scarcely be affirmed of pure simple earth."[79] According to Boerhaave's theory, the true decomposition of a metal produced "mercury," not earth.[80]

The differences between the theories of Boerhaave and Stahl, however, concealed a deeper divide in their philosophies for structuring the field of chemistry and setting its agenda for research. Most of Stahl's chemical texts were systematic monographs on broad chemical topics, such as fermentation, sulfur, salts, and assaying. Each espoused to explain chemical operations by invoking the action of a small number of chemical principles (such as phlogiston, water, earth, or the "universal acid"), which he argued comprised the examined chemical bodies.[81] This chemistry of principles, even though it originated in the seventeenth century, was the heart of Stahl's legacy to the chemists of the eighteenth century. Indeed, this approach was evident in Stahl's *Fundamenta Chymiae*, although Stahl had yet to develop his mature ideas in this text. Boerhaave, on the other hand, had adopted and refined his instrument approach to chemistry as a means of escaping the pitfalls he saw in the unexamined use of chemical principles. As stated earlier, by organizing his chemical lectures around the instruments, Boerhaave placed his emphasis on examining the natural mechanisms and "powers" that chemists manipulated in their operations with the aim of determining the true limits and capabilities of these operations. His arguments against the earth of metals was a good example of how Boerhaave applied this principle in practice. Of course, Boerhaave accepted into his chemistry entities analogous to Stahl's principles, but he did not build his theory of chemistry around them. He treated them as objects to be examined and explained though the instruments, not as principles used to explain a wide range of phenomena.

Chemical Affinity

Étienne-François Geoffroy's *table des rapports* published in 1718, initiated a flood of imitators, which made tables of chemical affinity an important

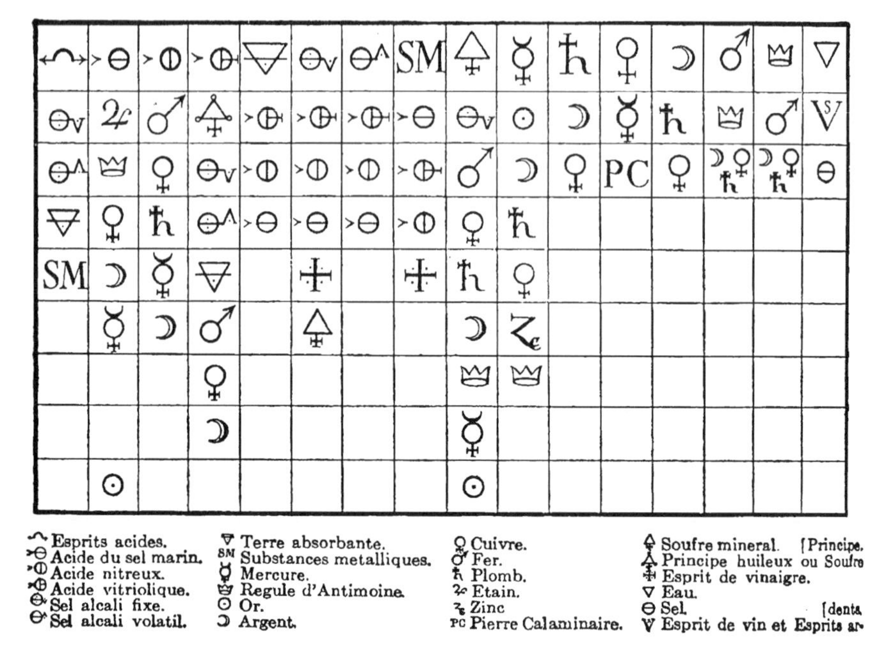

FIGURE 5. Geoffroy's *table des rapports*, 1718.

feature of later eighteenth-century chemistry.[82] When Boerhaave composed the *Elementa*, however, the future importance of affinity tables was as yet unrealized. Neither affinity tables nor the concept of "chemical affinity" became a dominant feature of chemistry until after 1750, initiated largely by the publication of Pierre-Joseph Macquer's *Elémens de chymie—Théorique* (1749).[83] Boerhaave did not discuss Geoffroy's table in the *Elementa*, even though he examined at various points the phenomenon ostensibly represented in the table: the displacement of chemical species in compounds, usually salts, by other species. One reason for this omission stems from Boerhaave's ordering of phenomena according to his instrument theory. He discussed this type of chemical action under the general heading of chemical menstrua, which included all instances of chemical species interacting through attraction. By approaching the problem through the concept of attraction, he strove to explain both the displacements represented by Geoffroy's table and also more complex interactions that Geoffroy could not represent in tabular form.

Étienne-François Geoffroy published the first table of chemical affinities, and as he stated in his *mémoire* on the subject, his aim was to illustrate the laws (*loix*) by which substances displaced others in compounds. He stated the general law of displacement as follows: "Every time that two substances which have some disposition to join themselves with one another find themselves a united whole; if a third [substance] happens by which has a greater *rapport* with one of the two, it [the third] unites with it and makes it release its grip on the other."[84] As indicated by this statement, Geoffroy did not employ the term "affinity" to describe the cause of displacement, preferring to use the more neutral term *rapport*. This move reflected the empirical tone of his *mémoire*. Geoffroy did not speculate regarding the mechanism of displacement and defined *rapport* only as the property that governed the "hidden movements" (*mouvements cachés*), which caused the mixing of bodies. He explained that the order of displacements he presented was based on experimental observations made by himself and other chemists.[85] In the table (see fig. 5) the symbol for an object-species or class of substances was listed at the top of a column, with each species that would combine with it listed below it. The species closer to the top of the column had the greater *rapport* with the object-species than those farther down the column. Higher substances would, theoretically, displace lower species from compounds with the object-species. By ordering these displacements in a table, Geoffroy introduced a new and useful method to order and explain a series of well-known phenomena.

Geoffroy's aims in constructing his table of *rapports* were simultaneously pedagogical, utilitarian, and reformatory. Geoffroy stated that he presented the order of *rapports* in tabular form for ease in use: so "one may see the different *rapports* at a glance." The ease in use had two underlying motives. The first was pedagogical. As Geoffroy argued, through the use of his table, the chemical student would quickly form a "just idea" of the *rapports* between various substances. Second, the table would be a useful tool for chemists in analyzing their operations and, by extension, revealing previous errors in interpretation: "chemists will find in it [i.e., the table] an easy method for discovering what happens to unravel several of their difficult operations, & what may result from the mixtures which they make with different compound bodies."[86] Put another way, the table could be used to predict the outcomes of chemical operations. Geoffroy's table was a product of and was designed to be useful for a specific type of chemical research. As Frederic Holmes has pointed out, Geoffroy's table grew out of the research tradition established by Wilhelm Homberg at the Académie royale des sciences in the early eighteenth century. Homberg and his younger colleagues Geoffroy and Louis Lemery (1677–1743) endeavored to find empirical evidence for their system of chemical principles, which they identified as salt, sulfur, mercury, phlegm, and earth.[87] Although only one principle (sulfur) was included in the table, it was evident that the table was supposed to aid in the study of chemical composition. The information embodied by the table assisted the researcher in identifying the products of his operations but not in understanding their mechanisms or the practical work to obtain those products.

Boerhaave too shared Geoffroy's pedagogical values and reforming spirit, but his scheme for ordering chemical phenomena had a broader scope than Geoffroy's. The serial displacements, which Geoffroy organized under the term *rapport*, were only a part of Boerhaave's more encompassing concept of chemical menstrua. In the *Elementa* he defined a menstruum as "a Body, which being applied to another . . . would so minutely divide it, that the Particles of the Solvent should be perfectly intermixed with those of the dissolved body." In the ensuing discussion, he classified Geoffroy's displacements as a special kind of menstruum in which the solvent united with one part of the solute.[88] He suggested that menstrua should be ordered according to their "manner of acting." Following his classification that he outlined in his instruments course, Boerhaave named four general types of menstrua based on their manner of acting: those achieved by mechanical action, by the "power of repulsion," by mutual attraction, and by a combination of the previous three.[89] Although Boerhaave discussed at length the importance of the mechanical action of "fire," the most important types of menstrua were those involving a "sociable attraction" between solvent and solute. These attractions were specific "powers" that arose between two types of bodies and could only be deduced through experience.[90] Because one could not predict these attractive powers universally, Boerhaave devoted most of his discussion of menstrua to examining the qualities of specific types of solvents: watery, oily, spirituous (i.e., alcohol), and saline (acid, alkali, and neutral salt). Under the aegis of "attraction," he encompassed a wide range of phenomena that extended far beyond solution chemistry. For example, he included under "aqueous menstrua" an examination of the manner in which alkali salts absorbed water from the air, eventually seeming to melt if left in the open. In a reductionist fashion, Boerhaave argued that there were two "powers" at work in this phenomenon: one that "attracts" water from the air to the salt, and one that dissolves the salt in the water.[91]

In sorting out the various attractive powers of menstrua, Boerhaave endeavored to show the complexity of phenomena involved. One of the primary criticisms of affinity tables during the eighteenth century was that the order of displacement could be altered by various factors, such as heat and the method of combining the substances. In fact, Geoffroy himself recognized that the displacements were "not always exact and precise," and as early as 1720, he published a second *mémoire* that addressed criticisms he received regarding the 1718 table.[92] Boerhaave ordered his menstrua according to type to encompass seemingly idiosyncratic phenomena into his account. He suggested that only when one had collected a complete and detailed account of the characteristics of individual types of menstrua could their behavior and mechanisms of acting be known with certainty. Under the heading of "fixed alkali" menstrua, he presented the order of displacements (similar to that reported on Geoffroy's table) for acids with fixed alkali. He reported that spirit of niter, sulfur, or vitriol would force an alkali compounded with vinegar to release the vinegar, and spirit of vitriol would displace any other acid compounded with alkali. Spirit of niter and spirit of salt, however, regardless of which acid was compounded with the alkali, displaced the other only partially, generating a quantity of aqua regia (spirits of niter and salt mixed) and precipitating a type of "nitrous salt" in which both acids combined with the alkali to some degree. This situation indicated to Boerhaave that there was not great difference between the two acids "with respect to their acid quality."[93] Later, in the section on acid menstrua, he described methods by which either spirit of niter or spirit of salt may be distilled with a salt from the other acid to form aqua regia. The significance of this mixed acid, as Boerhaave related, was that it dissolved gold, unlike either spirit of niter or spirit of salt alone. Thus, a chemist could create a new acid with completely different powers of dissolving by simply mixing two mineral acids together.[94] Aqua regia was a well-known and important chemical reagent, especially in assaying, but this acid did not appear on Geoffroy's table because it was not a simple substance. This consequence of the table's structure, along with the fact that Geoffroy ranked the displacement of acids, completely concealed the complex behavior of spirit of niter and spirit of salt to form aqua regia, as well as the specific powers of aqua regia as a solvent.[95]

Ultimately, Boerhaave did not embrace Geoffroy's *rapports* because they did not support Boerhaave's agenda for chemical reform. Boerhaave's approach to chemical menstrua was to collect and explain the characteristics of the various types so that one day their mechanisms might be perfectly understood. This project included the examination of complex and idiosyncratic phenomena, because he thought that these were key to understanding the activity of each specific type of menstruum. Geoffroy wished to organize chemical interactions, but by the very nature of the tabular form, he could not easily represent many of the complex phenomena, which drew Boerhaave's strict attention. On a more fundamental level, Geoffroy designed his table to assist chemists in studies of chemical composition. While Boerhaave was also concerned with composition, he focused his efforts on examining the mechanisms of chemical interaction, and the wide variety of phenomena organized under the rubric of chemical menstrua reflected this approach. In the end, however, chemists in the second half of the eighteenth century combined these two approaches. While they presented chemical displacements through tables, they also attributed the cause of these displacements to specific and relative attractions between types of substances.[96]

CONCLUSION

The *Elementa Chemiae* was the most important chemical textbook published during the first half of the eighteenth century because it helped to set the pedagogical and research agenda for chemistry. Rather than present recipes and techniques for performing standard chemical operations, Boerhaave encouraged his students and readers to engage in experimentation for the purpose of working out the theoretical principles of chemistry. The *Elementa* also provided a framework for conducting this research and organizing extant chemical phenomena—the instrument theory. Few chemical lecturers before Boerhaave discussed the chemical instruments in the manner that he did (Bohn and Stahl were exceptions), but this approach to understanding chemical operations gradually gained ascendancy. In France, for example, G. F. Rouelle (1703–1770), the supposed font of French Stahlianism, modeled his chemical lectures on the *Elementa* by discussing the five chemical instruments and rejecting previous (didactic) systems of chemical principles for the same reasons Boerhaave had.[97] At the same time in Scotland, Boerhaave's pedagogical structure was replicated in the chemical lectures of Andrew Plummer, William Cullen, and Joseph Black.[98] None of these lecturers followed Boerhaave verbatim in all matters, but each (following Boerhaave's process) integrated new ideas into his existing pedagogical structure to fit their own conclusions and new developments into the field. For example, each integrated a discussion of phlogiston in their lectures on fire, and all (except Plummer) displayed a version of Geoffroy's affinity table during their lectures on menstrua.[99] The significance of the *Elementa Chemiae*, then, was not necessarily the chemical theories that Boerhaave espoused (although his work on "fire" and "menstrua" was widely praised);

instead, it was the usefulness of Boerhaave's method as a framework for organizing chemical knowledge, presenting it in courses, and guiding new experimentation. The *Elementa* provided a reliable and philosophically informed structure for chemistry, but one that was malleable enough to incorporate new developments in the field.

Perhaps more significantly, Boerhaave employed his pedagogical method to redefine chemistry as a discipline. He portrayed chemistry as an academic art: a practical activity guided by "natural," theoretical principles. The goal of the academic chemist was to search out these principles experimentally. In this manner, Boerhaave pursued chemistry as an academic subject related to natural philosophy, in which experimental research was a priority. Although he did not initiate the idea that chemistry was a form of experimental philosophy, he codified and defined the program for experimental chemistry in the *Elementa*. Through this book, he helped to set the agenda for academic chemistry for the next forty years.

CHAPTER SEVEN

From Alchemy to Chemistry

Herman Boerhaave was a patient man. Louis de Jaucourt (1704–1779), a former student of the famous medical professor, recounted a story of his patience and self-control. According to Jaucourt, one day while Boerhaave was giving a lecture on medicine, one of the boys who assisted him with his chemical work entered the auditorium to stoke the fire in one of the chemical furnaces. The boy, however, was too hasty in his work and overturned the cupel inside the furnace. According to Jaucourt, Boerhaave at first turned red. Regaining his composure after a moment, he informed his auditors in Latin that "an operation lasting twenty years on lead has vanished in the blink of an eye." Boerhaave then turned to his anxious laboratory hand and said, presumably in Dutch, "I assure you, it is nothing; I was wrong to demand from you a perpetual attention that is beyond human capacity." After consoling the boy in this manner, Boerhaave returned to his lecture with the same countenance, Jaucourt asserted, as if he had lost an experiment of only a few hours.[1]

This story is one of the various exempla that Boerhaave's students and friends related about central aspects of his character: his even temper, self-control, and persistence. Another dimension to this story, however, relates specifically to the experiment itself. Boerhaave's overturned cupel contained one of the many experiments he performed that were designed to transmute metals. Jaucourt's story reveals that Boerhaave did not conduct these alchemical experiments in private but maintained them in the same space in which he conducted his lectures and in full view of his students. The chemical *laboratorium* at the University of Leiden was not intended to be a place for conducting research (in the modern sense) but was, like Leiden's *theatrum physicum*, a pedagogical space. Thus, Boerhaave conducted chemical experiments using the same furnaces and equipment that he employed for the chemical demonstrations performed for his courses.

Boerhaave began this work on metals when he became professor of chemistry in 1718. This project was a continuation of the experiments on transmutation and the composition of metals that he had largely abandoned in 1702, after he became a medical lecturer at Leiden. His agenda to reform chemistry and medicine by "expunging the errors" from these fields shaped his approach to this project. He worked to collect, evaluate, and test claims regarding the composition and properties of metals taken from a wide variety of sources, including alchemical treatises and other medical and philosophical texts that discussed transmutation. Boerhaave saw this work as part of his reform of chemistry, but one which was only tangentially related to his courses. He, in fact, completed most of this work (ca. 1729–35) after he had resigned as professor of chemistry. He eventually published many of the conclusions from his experiments (mostly failures) in a series of papers titled "De Mercurio Experimenta" that appeared in the *Philosophical Transactions of the Royal Society of London* and, in an abbreviated installment, in the *Mémoires de l'Académie royale des sciences*.

These papers show the extent to which Boerhaave's pedagogical method came to be a generic way of organizing experimental knowledge. He presented his work as a series of experiments, or rather demonstrations, which he composed according to the same demonstrational method that he employed in his instruments course and in the *Elementa Chemiae*. Each demonstration served to amplify a particular phenomenon under examination and was embedded in a literary form that guided the reader toward his interpretation of the question under scrutiny. Whereas in the instruments course this method of presenting demonstrations served the pedagogical function of guiding students to interpret chemical phenomena according to Boerhaave's system, now he deployed the same method to establish proof (or to disprove) claims about metals for an audience of natural philosophers and physicians. Taken as a whole, Boerhaave's project on metals supported in practice his statements from the *Elementa* that his method of discovery (fact collection and the extraction of principles through induction) and his pedagogical method (the orderly display of such principles) were two aspects of a unified approach to experimental chemistry.

Boerhaave's work on metals also demonstrates the persistence of "alchemical" knowledge within mainstream chemical practice in the early eighteenth century. The claims that Boerhaave tested in "De Mercurio Experimenta" were not antiquarian; most were drawn from relatively recent sources. At this time, many chemists considered the transmutation of metals to be a possibility, a belief supported by the theories that underlay their understanding of the chemistry of metals.[2] The redefinition of alchemy as a pseudoscience that dealt solely with gold making (*chrysopoeia*) and preternatural medicines (such as the elixir of life) was under way but not yet completed.[3] In fact, in his chemical lectures and in the *Elementa Chemiae*, Boerhaave taught that metals were compounds of mercury and sulfur, a version of the Geberian dyad theory that he had first encountered in his early alchemical experimentation.[4] Over time, however, he developed an increasing skepticism toward some alchemical claims and embarked on a program of testing the implications of the sulfur/mercury dyad. As a result, "De Mercurio Experimenta," written from 1733 to 1736, presented a much more critical view of alchemy than that found in the *Elementa*, composed from 1728 to 1729.

Ultimately, Boerhaave came to reject several methods for the transmutation of metals and, ultimately, came to doubt the sulfur/mercury dyad as the basis of their composition. This chapter examines this shift in Boerhaave's thinking and shows how the process of applying his pedagogical method to his alchemical research drove his increasing skepticism. I begin by describing his version of the sulfur/mercury dyad and showing how he presented this theory in his chemical lectures and in the *Elementa*. Then, I briefly describe some of his alchemical research projects based on extant experimental notes and examine his strategies and methodology for testing claims derived from the dyad theory. Last, I discuss his published conclusions from his alchemical work "De Mercurio Experimenta." Boerhaave formed his experiments (and often repeated experiments specifically to report them in print) to fit the demonstrational method he employed in his instruments course and in the *Elementa*. In the end, the method that he developed, shaped by his academic context and pedagogical concerns, led him to reject his belief regarding the composition of metals, which he had held for most of his career.

ALCHEMICAL WORK

When Boerhaave was appointed professor of chemistry in June 1718, his first order of business was to make an inventory and restore the university's chemical laboratory. This task proved daunting because the previous professor, Jacob le Mort, had made little effort to maintain the laboratory during his sixteen years as professor. Boerhaave spent much of his time from July through November 1718 supervising a project to enlarge and refurbish the laboratory, which now housed six new ovens.[5]

Although many of the investigations that Boerhaave undertook in this laboratory centered on commonly known chemical operations, much of his experimentation built upon alchemical studies he had suspended in 1702, when he became a lecturer in the medical faculty. His alchemical studies focused on the claims and theories of the mercurialist school. This school of alchemy, which I described in chapter 2, held that metals were compounds of a mercury and a sulfur principle, but that "mercury alone" was the root of the philosophers' stone. Boerhaave's earliest experiments in the Leiden laboratory tested mercurialist recipes to extract various "running" mercuries from metals and to "fix" small amounts of mercury into silver or gold. Finally, he continued his work on the "sophic" mercury of Eirenaeus Philalethes, eventually digesting this mercury with gold in an attempt to make the philosophers' stone. Boerhaave, however, was not interested in *chrysopoeia* as a means to generate wealth. Rather, he scrutinized the claims and theories of the mercurialist school that were then part of an accepted, if somewhat marginal, body of chemical knowledge and practice. He pursued this alchemical work in an orderly fashion and was not secretive about this research: his aim was to establish experimentally which processes had factual merit and which did not.

Boerhaave presented the mercurialist sulfur/mercury dyad in his chemistry courses and in the *Elementa Chemiae* as the proper theory through which to understand the behavior and properties of metals. He accepted the mercury/sulfur dyad as an established fact, primarily on the strength of the theory's endorsement by a legion of past mercurialists, such as "Geber," Robert Boyle, Joan Baptista van Helmont, and Wilhelm Homberg.[6] In his courses, Boerhaave followed the taxonomy of metals he found in the fourteenth-century text *Summa-Perfectionis*, attributed to "Geber." Here the seven traditional metals (gold, silver, mercury or quicksilver, copper, iron, tin, and lead) were understood according to a hierarchical arrangement based on their purity.[7] In his earliest lectures dating from 1702, he explained that metals grew in the earth from the combination of a pure, simple mercury and a subtle, "fixing" sulfur. Distinctions between the metals were based on the purity of their sulfur. The "best metals" (*metalla optima*), such as gold and silver, came from the purest sulfur and occupied the top of the hierarchy, whereas the "worst metals" (*metalla pessima*), such as lead and tin, came from sulfur having the most impurities of earth and salt.[8] By the time he composed the *Elementa Chemiae* (ca. 1728–29), however, he had deviated from this original scheme by positing that metals were composed of up to four types of substances. All metals contained a "most pure and simple" mercury and an "exceedingly subtle, pure, and simple" sulfur. All metals except gold also contained a "light" earth, which was specific to each type of metal and, due to its lesser density, accounted for the lower specific weights of the other metals compared to gold. Some of the base metals (lead, tin, and sometimes iron) also contained a greater or lesser degree of a "crude" sulfur, which Boerhaave identified as the scum (scoria) that these metals cast up when they were brought into fusion.[9]

Boerhaave's version of the mercury/sulfur dyad, however, was not simply an academic recasting of an old medieval theory. His theory of metals was an amalgamation of several mid- to late seventeenth-century views. He utilized his dyad theory to inform his own alchemical work, but its credibility (for Boerhaave) stemmed from Robert Boyle's advocacy of the theory. In his *Sceptical Chymist* and elsewhere, Boyle defended the mercury/sulfur dyad as the most suitable theory for explaining the composition of metals. Within the context of critiquing the various systems of chemical principles and elements, Boyle argued (going, it seems, against the claims of the *Summa Perfectionis*) that there were reliable accounts of chemists extracting material "sulfur" or "mercury" from metals by various processes.[10] Building on Boyle's argument, Boerhaave asserted in his chemistry lectures that the mercury and sulfur of metals were material substances, not abstract chemical principles. Metals were compounds of a material mercury and material sulfur related to but not identical with the "crude" (i.e., impure) quicksilver and mineral sulfur commonly found in the chemist's laboratory. In the *Elementa* Boerhaave asserted that any metal may theoretically be resolved into these "compounding elements," although the quantity and quality of mercury and sulfur produced will differ for each metal. Depending on the metal under analysis, he suggested applying one of several methods that used "mercury" (purified quicksilver), a "resuscitating salt" (alkali), or fire to effect the resolution of the metal.[11]

Boerhaave's initial work on the dyad theory did not involve the search for the philosophers' stone but rather focused on various processes related to "lesser" transmutation. These lesser transmutations (or *arcana minora*) involved the extraction or partial transmutation of metals by methods other than the "projection" of the philosophers' stone. Whereas the philosophers' stone (or *arcanum maius*) in traditional alchemical theory worked through the manipulation of a fermentive force or seed (*semen*) and effected "universal" transmutation, lesser transmutations manipulated this force only partially or imperfectly.[12] Many lesser transmutations derived directly from mercurialist theories of metals, and Boerhaave undertook the study of these processes for that reason. In testing the claims of the mercurialists, he sought to connect the empirical observations that he made in the laboratory with the mercurialist theories upon which the processes were based.

In winter 1720–21, Boerhaave initiated a series of alchemical experiments in the university laboratory that centered around the amalgamation and "digestion" of mercury with other metals. The aim of these experiments was to extract "running" mercuries—the internal, mercurial principle—from these metals. On November 28, 1720, Boerhaave placed a cupel containing one ounce of "the best English tin" into one of his furnaces, heated it to fusion, and added two ounces of quicksilver. As he cooled the resulting amalgam, it turned black, although it remained liquid. He gently heated the mass overnight, and the next day he removed a "black powder" (*nigerrimi pulveris*) from the amalgam by washing it with water. He reported in his notes that the washing left the amalgam looking "like silver." He again heated the mass overnight in a sand bath, and the next day he reported a "running mercury" in the bottom of the vessel. In another experiment that he began in March 1719, Boerhaave amalgamated a half pound of lead with a half pound batch of quicksilver, "purified with vinegar, salt, and pressed (with leather)." Again, the amalgam turned black, and he removed the blackness by washing the amalgam with water. Washing left the amalgam a silvery color. He again heated the amalgam and found mercury in the bottom of the vessel, which he called the "mercury of lead."[13]

To understand the significance of these operations, one must view them within the context of mercurialist theory in which Boerhaave worked. In these first amalgamation experiments, he was attempting to extract the pure, internal mercuries from tin and lead, and the methods he employed to do this derived from the practice of mercurialist alchemy. The mercuries extracted from metals were called "running" or "flowing" (*currens*) mercuries because they "ran" in the bottom of the vessel during their extraction from the metal. To extract the running mercury, Boerhaave prepared another type of mercury from ordinary quicksilver, which he described as *optimus*, a term that reflected the substance's purity. Such a mercury would have been distilled at least once, if not several times, washed with vinegar and salt, and then "pressed" (i.e., strained) through leather, and sometimes "digested" at moderate heat for extended periods of time. Within the mercurialists' theoretical framework, such purification drove the larger particles of the crude impurities to the surface of the mercurial mass and concentrated the more subtle, pure "mercury" in the center. Once the "sulfurous" impurities had been removed, the *optimus* mercury, acting like a "sophic" mercury through the aid of heat, could (in theory) penetrate the pores of the metal, releasing the metal's internal mercury.[14] The black powder produced during the amalgamation experiments was (theoretically) composed of the sulfurous and other "heterogeneous" impurities expelled though gentle heating (i.e., digestion). Once he had washed away the black powder, Boerhaave recorded that each amalgam then had a silvery color, a sign of its purity and an indication that a pure mercury could then be extracted from it.

Boerhaave experimented with other methods for extracting running mercuries from metals, but some of these experiments revealed inconsistencies among practitioners within the mercurialist school. In November and December 1723, he investigated a method that he attributed to Isaac Hollandus (fl. 1560–70s) to produce a running mercury from lead, using vinegar as the agent of extraction.[15] The theory behind the process was that the particles of the acidic salt of the vinegar were able to penetrate the particles of lead, releasing the lead's internal mercury.[16] After creating a mercury and lead amalgam, Boerhaave added a strong vinegar of wine to the amalgam. This action generated (after separation) a quantity of pure mercury, a "golden liquor," and some black powder. According to Hollandus's interpretation of this operation, the golden liquor was a solution of salt of lead, and the mercury was the original mercury from the amalgam. To produce the mercury of lead, one mixed and heated the black dregs in a fixed alkali solution, which (according to Hollandus) caused the running mercury contained in the dregs to precipitate. Boerhaave, however, found that he was able to convert some of this powder into mercury simply by heating it.[17] He was suspicious of this result, probably because of his earlier work in Philalethan alchemy. Philalethes argued that the black powder produced through the digestion of mercury was composed of its crude, "heterogeneous" parts, and that one could obtain a fluid mercury from the black powder though heating.[18] To test whether his mercury was the mercury of lead or Philalethes's crude mercury, Boerhaave repeated Hollandus's recipe twice during 1731–35, but he began with lead alone. In each case, no running mercury appeared after the lead/vinegar solution was digested. He then added fixed alkali salt to the solution, hoping that this would cause the mercury of lead to precipitate from the dregs, but he was not able to obtain any mercury.[19]

Boerhaave also examined a different theory that accounted for the behavior of the powders and dregs produced from mercury. Wilhelm Homberg posited that the chemical principle sulfur, fire, and light were each different forms of the same substance. According to Homberg's account, mercury when heated by fire was bombarded by sulfurous particles of various sizes and grades, some of which adhered to the particles of mercury. As sulfur began to cover the mercury particles, the body of the mercury empirically seemed to turn black and, as the sulfur coat became greater, red. The sulfur coat could be removed through the agitation caused by further heating, releasing the fluid mercury once more. Homberg developed this theory to account for the transmutations of mercury into gold and silver that he observed in alchemical processes called *minera perpetua* (perpetual mines). In these operations, successive digestions of mercury were employed to produce the red powder, which when collected and then heated strongly, produced fluid mercury and a small quantity of gold or silver. Homberg reasoned that during this operation a small amount of the most subtle particles of the sulfur penetrated into some of the mercury particles so that the fiery sulfur "fixed" the mercury, producing gold or silver.[20]

Boerhaave conducted experiments to test both the *minera perpetua* and Homberg's theory of the equivalence of sulfur and fire. He was introduced to Homberg's theory from its source; the two men had corresponded during Homberg's lifetime, and the latter sent Boerhaave a complete manuscript version of his "Essais de chimie," in which he outlined his theory.[21] To test Homberg's claims, Boerhaave conducted several different experiments. In 1718 he began a long-term experiment in which he digested a sample of purified quicksilver in an open vessel at moderate heat. He continued this digestion, having an assistant maintain the fire and checking on the experiment's progress a couple of time per week, for fifteen years.[22] In one of several shorter experiments he conducted in 1727, Boerhaave heated a sample of mercury to dryness, obtaining a red powder—the same red powder that Homberg had described in his "Essais de chimie." In notes written at the end of this experiment, Boerhaave expressed his uncertainty regarding the nature and qualities of this powder: "What is this powder? Examine through new experiments, especially cohobating mercury with the powder itself."[23] Boerhaave finally began these "cohobating" experiments in September 1731, starting with a seventeen-ounce sample of pure quicksilver. During the next twenty-six months, he distilled the quicksilver more than five hundred times, eventually collecting just over seven ounces of the red powder.[24] On December 11, 1733, Boerhaave heated the red powder alone in an alembic, as he stated, "so I may know if any fixed metals may be contained in it." He converted all of the red powder, save fifteen grains, into "the purest mercury." He heated the remaining powder in a stronger fire, but he could neither reconvert the powder back into mercury nor obtain any solid metal from it.[25]

After the *Elementa Chemiae* was published, Boerhaave resumed his efforts to prepare the *arcana maius*, the philosophers' stone. He began with another attempt to make a "sophic" mercury according to the recipe of Eirenaeus Philalethes. On March 22, 1732, Boerhaave recorded in experimental notes that he obtained twelve ounces of "[mercurius] Suchtenii," a reference to Alexander von Suchten (ca. 1520–75), whom Boerhaave had identified as the originator of Philalethes's recipe for sophic mercury.[26] To this mercury he added two and one-half ounces of pure silver and one and one-half ounces of *reguli antimonii martialis*, which he had prepared from November 11 through December 12, 1731. He digested the mass for three days, washed away the "blackness" that coated the mass, and distilled the mercury out the amalgam. Through August 12, 1732, he repeated this series of operations eight more times. Following Philalethes's nomenclature, he called the resulting mercury, distilled nine times, the "ninth eagle."[27] Throughout the process of "taking" the eagles, Boerhaave carefully weighed his starting materials and products, and he identified the composition of all amalgams and by-products. Inspired by his earlier investigations of the various powders generated during the digestion of mercury, he was especially interested in the "blackness" generated during these processes and resolved to determine its composition. Boerhaave distilled the black powder and reported that he recovered "pure regulus" (of antimony) covered with "glass" and "cinders."[28]

At this point Boerhaave had followed Philalethes's recipe for "sophic" mercury to its completion. The only step that remained to create the philosophers' stone was to digest the sophic mercury in gold. On January 22, 1734, he obtained two ounces of "the most pure" gold, which he cut into "thin fragments." In a mortar, he added to this gold six ounces of "mercurii Suchteniani IX aquilarum." The gold and mercury easily amalgamated, and he placed the mortar over a furnace in order to dry its contents completely. He then placed the dried amalgam in a "glass egg" (*ovum vitreum*), which he sealed with paper and heated in a sand bath at moderate heat for five days. Satisfied that all of the water was expelled (and would not break the glass), he sealed the glass "hermetically" and placed the "egg" back into the sand bath. Boerhaave recorded that after five months of heating, the contents of the egg had turned black, and after twenty more days, the mass began to turn white, which it did completely by October 1, 1734. Boerhaave continued to heat the amalgam until September 3, 1735, at which time it had remained white for eleven months with no visible change.[29]

While he was digesting the first amalgam of gold and sophic mercury, Boerhaave made a second sophic mercury according to Philalethes's method. On July 5, 1734, he began the process of amalgamating, digesting, and sublimating mercury with silver and *regulus antimonii martialis*. On March 12, 1735, Boerhaave took the "twelfth eagle." Just as he had after making his earlier "sophic" mercury, he collected, weighted, and analyzed the black dregs produced during the operations. On March 20, 1735, he amalgamated three ounces of gold with thirteen ounces of mercury of the "twelfth eagle." He placed the mass inside a "philosophical glass" (*vitrum philosophicum*) and heated it in a sand bath until September 3 (the same time at which he ended the first experiment). Boerhaave did not record any observations on this amalgam or comment on the experiment's success.[30] Although he did not record his assessment of the experiment, one might infer from the context that he did not believe that he had succeeded in making the philosophers' stone in either of his two trials. He did not attempt Philalethes's process again.

Boerhaave's alchemical career in the laboratory was one marked by failures. In seventeen years of experimentation, he was scarcely able to bring any of the mercurialist recipes to fruition in a way that satisfied his growing skepticism. In some processes, such as his *minera perpetua* experiments, he was unable to obtain any of the results claimed by mercurialist alchemy. He then submitted his seemingly successful experiments, such as the running mercury extractions and Philalethes's method for sophic mercury, to analytical scrutiny. By carefully analyzing each product and following each substance through the various operations in each process, he revealed inconsistencies and misinterpretations in these processes as well. Although he did not possess the philosophers' stone, Boerhaave had collected a wealth of firsthand information on mercurialist alchemy. In the next section, we see how he communicated this knowledge to his peers and students.

MERCURY EXPERIMENTS

Boerhaave published the conclusions of his alchemical experiments in a series of three papers, each titled "De Mercurio Experimenta." Despite their innocuous title, these papers presented a devastating experimental critique of mercurialist alchemy and reflected his growing skepticism regarding the possibility of metallic transmutation. The effectiveness of his critique derived largely from his ability to impose a rational order on his various experiments, directing his presentation toward rebuking specific claims. For this process, he employed the same demonstrational method (i.e., a description of a performed *experimentum* followed by interpretive *corollaria*) that he used in the *Elementa* to fashion his arguments. Ultimately, "De Mercurio Experimenta" illustrated Boerhaave's contention that his pedagogical method was both the most efficient way of presenting a body of facts and the best way to establish factual or theoretical claims based on those facts. In this case, the very process of organizing his previous experimental results according to his method forced Boerhaave to resolve inconsistencies and ambiguities in his understanding of metals and challenge the view he had held previously. By bringing his previous experimental failures (from the mercurialist perspective) into acute focus, the act of methodizing his alchemical work encouraged his skepticism regarding the transmutation of metals and the mercurialist theories that supported this belief.

When Boerhaave wrote the *Elementa Chemiae* in 1728–29, his discussion of "alchemy" in that book resembled an apology for the art rather than a criticism. He presented his comments on alchemy in the section of the text where he outlined the usefulness of chemistry for other arts and sciences. Like the previous arts he discussed, he claimed to present "a few simple observations concerning the usefulness of the chemical Art in Alchemy."[31] Here, he completely ignored the growing collection of literature by leading chemists, which strove to debunk certain alchemical operations and portray alchemists as charlatans.[32] Instead, Boerhaave called them natural philosophers and asserted that no one had "more intimately examined, and evidently explain'd the nature of Bodies, and the effects they are capable of producing."[33]

For Boerhaave, the problem with the alchemists came during the execution and testing of their theories. Invoking the same mistake that in his oration *De Chemia suos Errores Expurgante* he had attributed to all chemists, he argued that the alchemists fell into error by interpreting their speculations as facts without conducting proper experimental investigations. He supported this claim by citing an unlikely source, Alexander von Suchten. As Boerhaave recounted, Suchten stated at the end of "one of his Treatises of Antimony" and "after he had tried a great many things to no purpose" that "all the Philosophers . . . died before they had brought their speculations to proper issue."[34] This realization, however, did not dissuade Boerhaave from continuing his enthusiasm for alchemy. Despite the fact the many of the alchemists' claims had not been proven true, he maintained that in searching for the philosophers' stone they had discovered many useful facts and processes. He recounted a parable from Francis Bacon to illustrate this point: "And the great lord *Verulam* very properly compares them [i.e., alchemists] to a Father, who when he was dying, made his lazy sons believe, that in a certain field he had hid a treasure; upon which, as soon as he was dead, they went to digging, in hopes of finding riches, but were baulk'd in their expectations, tho' the service they did the Land abundantly recompenc'd their trouble." As an example of this principle, Boerhaave pointed to the recent discovery of phosphorus by Daniel Crafft and Johann Kunkel, the thought of which would previously have seemed "incredible." He concluded his discussion of alchemy with a plea for moderation, arguing that one should be neither too credulous nor too skeptical: "It is the business of the wise man to try everything, and abide by that which he finds to be true; nor ever to prescribe limits to the power of the omnipotent Governor of the universe, or the natural beings which he has created."[35]

Boerhaave continued his experimentation after the publication of the *Elementa*, but it was only through a little prodding that he eventually published any of these results. The event that led to publication was Boerhaave's election to the Royal Society of London on April 30, 1730. Boerhaave soon after received a letter from Cromwell Mortimer (1689–1752), a former student who was soon (in November 1730) to be elected secretary of the society. In this letter, Mortimer encouraged his former mentor to submit something to him that he might publish in the Royal Society's journal, *Philosophical Transactions*. After an extended illness, which left him incapacitated from September 1731 through February 1732, Boerhaave finally agreed to send a report on the "remarkable properties of quicksilver examined though the most laborious experiments."[36] With this in mind, he redoubled his efforts in the chemical laboratory, initiating several new experiments with the specific aim of presenting them in his account for the *Philosophical Transactions*. He sent off his first paper, titled simply "De Mercurio Experimenta," to Mortimer (now secretary of the Royal Society) in February 1734. The paper was included in the December 1733–34 edition of the *Philosophical Transactions*. Boerhaave sent two more installments in March 1737, which appeared in the October and December 1736–37 editions of the *Philosophical Transactions*.[37]

Boerhaave's papers on mercury contained a sustained critique of alchemy. As he did in the *Elementa*, Boerhaave demarcated between the "alchemists" and the more modern "chemists." In contrast with his discussion of alchemists in the *Elementa*, however, he adopted a critical tone from the outset. He began the first installment of "De Mercurio Experimenta" by attacking the communication practices of the alchemists, contrasted with those of the practitioners of *chemia*. He argued that the chemists had devised a sure way in which the "powers" (*vires*) of bodies may be established with certainty. These "authors," he continued, arrived at "clear propositions" as long as they remained true to their methods. The alchemists, in search of their "most ancient prize" (i.e., the philosophers' stone), had also used this method: "No group truly applied so much bitter and tenacious work dedicated to investigations of natural, mortal things or so much excessive labor by turning over the various ways of exploring matter as the alchemists."[38] The problem with the alchemists rested in their secretive and misleading modes of communication. Boerhaave asserted that they revealed their "easy" (*facile*) operations, yet concealed information regarding their most prized and difficult operations. By making "rich promises with serious words," but leaving the "thing itself in dark obscurity," they have eroded the confidence of wise men in their claims, who "drive them out as mad, fabulous, and empty liars."[39] Here, Boerhaave finally addressed the recent spate of literature that portrayed alchemists as charlatans, but he did not dismiss the alchemists outright as others had. While condemning their secretive practices, he suggested that the veil of obscurity be lifted and the alchemists' claims put to the test. As he asserted, "my chemical work was single-minded in scrutinizing and disclosing the writings of the alchemists for all to see."[40]

Boerhaave reiterated the alchemists' theory that metals grew in the earth from seeds and explained how this theory related to the alchemists' methods of gold making. He posited that according to a "written law of the underworld" (*per legem subterraneis scriptam*), four conditions—a living seed, suitable "nourishment" (*alimenta*), a proper matrix, and proper heat—were necessary for metallic growth, and as long as these four were present, gold would grow "in a material like itself" (*in materiem sibi similem*). This material, Boerhaave argued, was quicksilver, and the seed or "power of metallification" (*vis metallisica*) was sulfur.[41] He then revealed the core theory of the mercurialist school of alchemy by hinting that, theoretically, the mercury might be separated from its sulfur and purified, providing a suitable, artificial medium for growing the seeds of metals. If this mercury were to be "cleansed by art," it would be "fluid, metallic, most ponderous, simple and never divisible into parts by art or nature." He further asserted that in this mercury, "gold itself melts, is heated, and matured."[42] This, of course, was a description of the mercurialists' sophic mercury. Boerhaave then related how he himself had endeavored to learn how this mercury may be prepared by art.

When Boerhaave came to presenting the experiments, he structured the text in accordance with the demonstrational method he had used in the *Elementa Chemiae*. Following pedagogical principles, his *experimenta* were not, in fact, experiments in the sense that their outcome was in doubt but were demonstrations in the sense that they represented the usual outcome of the operations they described. To organize the information he needed to convey, Boerhaave divided the description of each numbered experiment into three parts: *operatio*, *effectus*, and *corollaria*.[43] As one may surmise from their names, the *operatio* described the experiment itself, the *effectus* described the results and products of the experiment, and the *corollaria* enumerated the factual points or theoretical conclusions that Boerhaave deduced from the experiment alone or in conjunction with previous experiments. As in his instruments course and textbook, this method aimed to present the inferences from experiment to a claim under scrutiny in a clear, orderly, and unambiguous manner.

The accounts in "De Mercurio Experimenta" described actual experiments that Boerhaave had performed, but they were highly sanitized depictions that often he had worked out through several experimental trials. He omitted what he considered to be superfluous empirical or operational details in order to present a clean account of the phenomena under scrutiny. When he had conducted an experiment several times, he chose the best result as his exemplar. He performed several of the experiments that he described in the text specifically for publication, either repeating an older experiment to generate a clean trial or designing a new setup to elucidate a particular theoretical point. Although he had devised many of the experiments in "De Mercurio Experimenta" before he had composed the text, the very act of composition and of shaping his claims into a suitable presentation often necessitated additional experimental work. This process was similar to the way in which he constructed his demonstrations for the instruments course (described in chap. 5).

The first experiment that Boerhaave presented was one of the later ones he performed. It centered on shaking common quicksilver for an extended period of time in order to generate a black powder. He first reported in the *operatio* section how he cleaned the mercury by rubbing it with water and sea salt and then distilling it. He placed the mercury in a vessel made from "green German glass" and heated the vessel in a sand bath until he was certain that all of the water from the washing had been removed. Boerhaave then placed the mercury in a "flagon" (*lagena*), sealed the opening of the vessel, and placed the vessel in a small box, through which the neck of the flagon protruded slightly. He affixed this box "to the striking arm of a fuller's hammer," which shook and concussed the mercury constantly from March 1 through November 11, 1732.[44] In the *effectus* section of the experiment, Boerhaave recounted how he recovered copious amounts of black powder as he pressed the mercury through leather. Finally, as *corollaria* he listed five statements, each of which indicated that the black powder was generated out of pure mercury through shaking alone.[45]

This first experiment was one that Boerhaave performed specifically for publication. During the 1720s, he had undertaken several variations on this experiment, which he mentioned in several places in the *Elementa*, attributing the discovery of this phenomenon to Wilhelm Homberg.[46] Boerhaave's conclusions were not remarkable on their own account, except for the fact that he had struggled for several years to determine the nature of the black powder. He performed this experiment precisely to show to his readers in an unambiguous manner what he had labored to discover: that the black powder was a form of mercury that could be generated without combining the quicksilver from which it came with any other (known) chemical entity. This experiment was not originally a part of Boerhaave's own process of discovery. Its function in "De Mercurio Experimenta" was rhetorical and pedagogical: the simplest way to establish a crucial fact. Having conveyed that the black powder came from mercury alone, he built upon this simple foundation a chain of subsequent experiments in which he purported to show that other claims regarding the composition of the black powder were false. By demonstrating the errors of these other claims, Boerhaave could then undermine the theoretical claims that were associated with the black powder's composition and, by extension, the chemical properties of mercury.

The next eleven operations that Boerhaave discussed involved generating powders in mercury through distillation. Although he did not give the dates of these experiments in "De Mercurio Experimenta," by searching his experimental notebooks one can deduce that they all took place between September 10, 1731, and November 27, 1733, indicating that Boerhaave also performed these experiments with publication in mind. The notes also imply that they constituted one continuous chain of investigation, but he presented them as eleven discrete experiments.[47] As described in his paper, he began with a sample of quicksilver, which he purified as described in the first experiment and distilled sixty-one times. He reported that five drams of red powder were produced, but he set this powder aside for the moment. He then took the mercury and submitted it to the same "shaking" that he described in the first experiment. He reported that he obtained the same black powder, which in the next experiment he reconverted back to fluid mercury by subjecting it to "great heat." Boerhaave asserted in the *corollaria* of these experiments that he had demonstrated that even highly purified mercury produced the back powder upon shaking, and that he could convert the powder back to mercury through heating. Now, he concentrated his efforts on the red powder. As I described earlier in this chapter, Boerhaave performed 511 distillations on the original eighteen-ounce sample of mercury. From this process, he obtained just over two ounces of red powder, of which he converted (or lost) all back into fluid mercury through heating, save fifteen grains, which, he stated, remained "fixed."[48]

Although Boerhaave did not state this overtly in the "De Mercurio Experimenta," he was, in fact, testing the alchemical *minera perpetua* and the theory of its mechanism devised by Wilhelm Homberg. Recall that Homberg had posited that fire was the same substance as alchemical sulfur, and by heating or distilling common mercury, one could "fix" small amounts of mercury, transmuting it into a metal, ideally gold or silver.[49] Boerhaave produced fifteen grains of "fixed" mercury through his distillations, but this mass did not appear to be metallic. To verify that this was the case, he sent the fixed mercury to an assayer in Amsterdam to be tested. The assayer reported that after testing part of the mass with lead (i.e., cupellation), he determined the sample contained no gold or silver.[50] After relating the results of the test, Boerhaave presented a set of fifteen *corollaria* that summarized the conclusions from the first dozen experiments. The primary conclusion was that no mercury was "fixed" into gold or silver, regardless of how long it stayed in the fire, how strong the fire was, or how many times it was distilled. From this empirical conclusion, he then attacked Homberg's theory that supported the fixing of mercury in this manner. He argued that "it does not appear through these experiments" that mercury and fire may form metals. "Therefore, fire . . . is not demonstrated to be the Sulfur of the Philosophers, fixing mercury into metals."[51] The "sophic sulfur," Boerhaave contended, was something else. He concluded this series of experiments on an acerbic note, calling the notion that gold and silver could be synthesized from mercury an "easy promise" made by "the ignorant and those devoted to imagination": "Safe from the false writings and prescriptions of the Sophists who promise such things from mercury and fire in the space of a short time, a few months, I in several years truly did not even begin to conceive the first or least indication of success."[52] These last remarks illustrated Boerhaave's growing disbelief and shrinking patience with mercurialist alchemy.

One of Boerhaave's *experimenta* was not an "experiment" at all but rather a novel comparison of facts presented directly to refute a theoretical claim. Nevertheless, he presented these facts and their interpretation in the same demonstrational form as the other *experimenta*. The fifteenth and last experiment in the first installment of "De Mercurio Experimenta" presented an argument against a common belief in mercurialist alchemy. Boerhaave began: "Geber writes that pure mercury is as ponderous as gold." By "pure" mercury, Boerhaave (and Geber) meant mercury that had been "philosophically" purified of all of its crude, sulfurous, and heterogeneous particles, leaving only the homogeneous, "sophic" mercury, which (according to the theory) was the material foundation of gold.[53] According to Boerhaave, one made such a mercury by "separating the lighter and mutating parts" from ordinary quicksilver. In Geberian alchemy, this task was usually accomplished through successive sublimations (or distillations) of a given sample of mercury.[54] Boerhaave reasoned that as the mercury was purified through distillation, and thus made more like gold, its weight and density should gradually increase, approaching those of gold. To test this claim, he measured the specific weight (*examinando hydrostatice in aqua pluva*) of gold and compared it to the specific weights of various samples of mercury from his own work. The specific weight of pure gold, he reported, was 19 119/500, whereas the specific weight of store-bought mercury, once distilled, was 13 57/100. He also reported the specific weights of mercury that he had amalgamated and recovered from gold (13 55/100), silver (13 58/100), and lead (13 55/100). Finally, he reported that the specific weight of the mercury that he had distilled 511 times was 14 11/100.[55]

In the *corollaria* section of this *experimentum*, Boerhaave presented these facts to question two mercurialist theories of "sophic" mercury. For the first, he examined the specific weights of the mercuries made from the amalgamations with other metals. Because the mercury extracted from the gold and lead amalgams had slightly lower specific weights than common mercury, he argued that "if purified [*defoecatus*] mercury is made lighter [*levior*], then the most purified [mercury] is rendered through [amalgamation with] gold and lead." He stated that these results seemed to support the claims of Philalethan alchemy: "One expects the same by the art of Suchten and Philalethes."[56] Boerhaave interpreted the slightly lower specific weight of the amalgamated mercury as a possible indication that the gold absorbed a fraction of the heavier (i.e., sophic) parts of mercury into itself. In a later corollary he asked, "Does mercury deposit its most weighty part in gold? Is this deposited part the seed of gold?"[57] In Boerhaave's mind, this phenomenon was predicted by Philalethes's theory of transmutation, in which at the finals stages of the transmutation process, the subtle, "sophic" mercury, which Philalethes also called the "seed of gold," entered into a union with the *semina* contained within the gold itself. This union was one of like with like, that is, subtle mercurial particles uniting *per minima* with the subtle parts of gold.[58] Boerhaave, however, identified a discrepancy between this theory and his observed results. The mercury he obtained from the lead amalgam seemed to absorb a heavy fraction of mercury just like the gold did. This phenomenon was not predicted by Philalethes, because in mercurialist theories of the composition of metals, lead was composed primarily of gross particles and should not be able to extract the subtle parts from mercury. Characteristically, Boerhaave simply pointed out the discrepancy without comment. In the same set of *corollaria*, he critiqued the Geberian theory of sophic mercury as well. Recall that in Geber's process, successive distillations were supposed to bring mercury closer in nature and in weight to gold. As Boerhaave related, after 511 distillations the specific weight of his mercury was nowhere near that of gold. He again asked, "would mercury by continued work finally be able to be made as dense as gold?" He answered skeptically, "Suitable judges may examine it."[59]

Boerhaave did not submit the second and third installments of "De Mercurio Experimenta" until early 1737, primarily because he did not complete all the experiments that he described in these papers until 1736. Most of the experiments described in the second paper covered familiar ground. The first two involved "digesting" mercury for an extended period of time to determine if any of the mercury became "fixed" into a metal. Boerhaave maintained one sample of mercury at "over" 100°F for fifteen and a half years (November 15, 1718, through May 23, 1734,) in an open vessel. This experiment was the first he included in his papers that (for obvious reasons) he did not repeat for publication. He did perform another related experiment, in which he heated a sample of mercury in a closed vessel for seven months (December 6, 1732, through July 8, 1733). These processes were slightly different versions of the *minera perpetua*, which Boerhaave had examined in the first paper. Like the earlier paper, he arrived at the same conclusions here: no mercury was "fixed" into precious metal, and alchemical "sulfur" did not seem to be the matter of fire.[60]

The remaining five experiments all concerned methods of extracting "running" mercuries from metals, primarily lead and tin. Boerhaave had attempted some form of each of these experiments during the 1720s, but as in his previous *experimenta,* he conducted new experiments for public presentation. For example, the first of the running mercury experiments concerned the extraction of "mercury of lead" using a "resuscitating" alkali. For almost forty years Boerhaave had pondered the claim that alkalis could dissolve the metallic, "fixing" sulfur of some metals, releasing their internal mercury. He first experimented on this idea in 1696 and then again in the 1720s. In the *Elementa,* he expressed doubts about the notion, pointing out that in most of the recipes for extracting running mercuries with alkali, one first calcined mercury or a metal with acids and recovered the "pristine" mercury when the alkali was added. Thus, the action of the alkali seemed to be on the acid, which had dissolved the metal, and not on its fixing sulfur.[61] In "De Mercurio Experimenta," Boerhaave presented a definitive experiment and commentary on the subject. He began by citing Joan Baptista van Helmont and J. J. Becher as authorities supporting this claim. He explained that he took a sample of white lead, dissolved it in spirit of niter (a strong acid), and then added sal ammoniac (an alkali) to the solution. He recovered a the white precipitate from the solution, mixed this matter with lixivial (i.e., alkali) salts, and calcined the whole mass in the furnace. He then "digested" the mass at 96°F for twenty-four months (February 1732 to February 1734), stopping several times to attempt to distill mercury out of the mass, although none was forthcoming. In the scholium following the experiment, Boerhaave related that several processes involving "resuscitating" alkali existed, but he also declared that he was never able to achieve successfully what the recipes claimed.[62]

In the third paper of "De Mercurio Experimenta," Boerhaave described a series of experiments in which he distilled mercury with gold. As the introduction to this third paper revealed, he considered these experiments to be a test of the action of the sophic mercury. He began the introduction of these experiments with yet another description of sophic mercury. He stated that the "ancient alchemists" claimed that when the alien "contagions" were purged from mercury, rendering it "simple," that it became immutable but also penetrating, able to dissolve other bodies without being changed itself. This mercury had wondrous powers and uses: "[It] digests crude metals, perfects base ones, and reduces whatever other bodies are at hand, resolving and converting them to a radical humidity; thus, [this] principle instrument for the secret recipes of medicine and the hermetic arts is rightly praised."[63] Boerhaave, however, citing the earlier installments of "De Mercurio Experimenta," argued that he had disproved the claims of the "hermetics" regarding the synthesis of this mercury. They claimed that mercury may be purified through distillation, but he had shown that the supposed impurities removed through distillation (i.e., the black powder) were able to be restored back into fluid mercury.[64] Now he was going to test the principle through which the sophic mercury worked. He explained that gold and silver were similar in nature to this "pure" mercury, "whether by birth or material." On the basis of this similarity, the alchemists wrongly believed that if ordinary quicksilver and a precious metal were amalgamated and then the mercury was distilled off, the metal would draw the purer part of mercury into itself, separating it from the "dirty" parts.[65] Boerhaave stated that he had studied this claim through experiments and that what he related in the paper were only those things that he found to be substantiated.

The experiment through which Boerhaave examined this claim entailed successive digestions and distillations of a mercury and gold amalgam. He had, in fact, investigated this claim since 1720, when he began experimenting on both gold and silver amalgams with mercury. In various experiments, some lasting until 1736, he attempted to "fix" mercury in gold or silver, both by "digesting" the amalgams and by successive distillation. In one case he digested an amalgam of gold and mercury at "moderate" heat for eight years (1720–27), and in another he amalgamated and distilled a sample of gold and mercury 389 times (ca. 1728–29).[66] As with previous topics, however, Boerhaave conducted a new experiment to act as the exemplar to appear in the published results. On July 1, 1734, he began this new experiment, amalgamating 2 ounces of gold in 20 ounces of mercury and then distilling off the mercury. He repeated this process 727 times over the next seventeen months.[67] On July 7, 1735, he began a second series of distillation experiments, this time starting with 2 1/2 ounces of gold and 25 ounces of mercury. He probably initiated this second series of experiments because he had lost a great deal of material (through handling, leakage, etc.) from the first experiments during the course of the operations. By the time he began the second set of distillations, he only had 7 ounces and 3 drams of mercury remaining (after 500 distillations) out of 20 ounces at the start. He was much more careful in the second series of experiments, which he described in the published account. Ultimately, he amalgamated and distilled this second amalgam of gold and mercury 877 times.[68]

Boerhaave analyzed and interpreted these experiments following the same methods he had employed in earlier experiments. He reported that the successive amalgamations of mercury and gold produced a brown powder, which he found that he could convert back into fluid mercury with strong heat. He commissioned his fellow professor, Willem 'sGravesande (1688–1742), to perform a hydrostatic analysis on the recovered mercury, which yielded a specific weight of 13½. In the *corollaria*, Boerhaave stated that this result indicated that the density of the mercury had not changed, and therefore no "light parts were liberated from it."[69] Thus, he asserted that the "fire" and "gold" did not separate "sulfur" or "feces" from the mercury, but rather the brown powder was yet another "species" of mercury. The final conclusion of Boerhaave's paper stated that mercury was not fixed by fire and gold as the alchemists had claimed, and he doubted that mercury may be concreted or fixed to augment or generate metals. Metals that appear to be mutated by fire, as Boerhaave discovered, were "stable": they could always be converted back into the original metal.[70]

CONCLUSION

My examination of "De Mercurio Experimenta" indicates that Boerhaave developed an increasingly skeptical attitude toward mercurialist alchemy during the composition of these papers. None of the experiments that he published supported the claims that he investigated. Around this same time (ca. 1733–35), Boerhaave also brought his experiments on Philalethes's recipe for the philosophers' stone to a close, which was seemingly another failure. These events reasonably explain his overtly negative tone regarding mercurialist alchemy in "De Mercurio Experimenta," as compared with other, more moderate statements on alchemy in the *Elementa Chemiae*. To say that Boerhaave's alchemical project was a complete failure, however, would be to miss the point. This work, especially within the context of its preparation for publication, must viewed as another example of his methodology of testing chemical claims in order to "expunge chemistry's errors." Many of the claims that he tested derived from recent sources and formed a reasonable, contemporary body of chemical theory, which explained the composition and properties of metals. Boerhaave, in fact, had taught this mercurialist theory in his chemistry courses for almost thirty years. What he accomplished in "De Mercurio Experimenta" was to *demonstrate*, using the same skeptical attitude and method of presentation from his chemistry courses, that mercurialist theories were, in the practical sense, false. Thus, the mercurialist theories of metals, including his own, could be discarded as another error removed from chemistry.

One would be mistaken, however, to assume that Boerhaave rejected all alchemical claims completely. Just as he was becoming convinced of the vacuity of mercurialist claims, he developed an eager interest in another school of thought on metals, which centered on a pre-metallic substance called "Guhr." In a series of letters (ca. 1732–35) to his friend and former student Joannes Baptista Bassand (1680–1742), Boerhaave described his growing enthusiasm for obtaining and investigating samples of "Guhr." He wondered if this material might provide a replacement for the sulfur/mercury theory of metals. As Boerhaave related, "German miners" (through the conduits of Agricola and Van Helmont) described this substance as an "oily fluid, as dense as fat, but usually yellow-green in color," which solidified to form metals. He contended that this substance might hold the key to a new theory for the composition of metals in which, against the assertions of Homberg and Boyle, the prime matter of metals would resemble vitriol (*vitriolus*) and not quicksilver.[71] Boerhaave, however, never brought his interest beyond the realm of speculation, and by 1736 he retired from chemical experimentation. Nevertheless, this example demonstrated how he remained open to all manner of chemical claims throughout his life and rejected only those that seemed futile in the laboratory. He retained the hope that there might be some truth to earlier theories of metals and their generation, even if he never found it.

Boerhaave's rejection of mercurialist alchemy in "De Mercurio Experimenta" also testified to his faith in his method. He had devoted almost twenty-five years of his life to alchemical experimentation and had published a moderately positive view of the subject in his textbook. Ultimately, however, he adhered to the pedagogical values that he had preached for his entire career. For Boerhaave, the ordering of empirical facts to generate theoretical conclusions according to a method reflected the logical structure of knowledge itself. In his mercurialist experiments, the principles he employed for presenting knowledge in an effective and orderly way to students were the same as those for presenting a coherent, philosophical argument. Boerhaave's method clarified his critique of mercurialist claims by allowing him to arrange experimental evidence in a manner that compelled his readers to doubt the veracity of those claims. Unlike the lying alchemists who inhabited the popular literature of the eighteenth century, Boerhaave methodized his alchemical failures into an exemplar of the experimental philosophy that grounded the proper practice of chemistry.

CONCLUSION

Boerhaave's Legacy

In April 1738, the Leiden curators dispatched the professor of botany Adriaan van Royen (1704–1779) to visit the ailing Herman Boerhaave, who had discontinued his medical teaching. Van Royen was charged with the task of getting his mentor's advice about finding a replacement to teach his courses. Ultimately, Boerhaave suggested that by dividing his duties among his former students already teaching on the medical faculty, Leiden would be able to maintain the progress made during his tenure. In an unusually candid statement, he listed what he thought his main achievements were. First among them was the fact that Leiden attracted many more students than the other Dutch universities. He was pleased that the "Stahlian method," Cartesianism, and "fantastical chemistry" (*phantastique Chemische*) had each fallen into disrepute, and he asserted that clinical instruction, which was reinstituted in 1714, must be maintained. In the end, he argued that his former students would be the best men to preserve these achievements as long as they followed the principles found in his books and teaching. They possessed, he asserted, a sound knowledge of medicine and a sound method of teaching.[1]

This episode from the end of Boerhaave's life reveals that he saw the reform of medicine and chemistry, especially his campaigns against what he took to be erroneous approaches, as his primary achievement. First and foremost was the rejection of "Cartesian" medicine, which Boerhaave defined as a rationalist method of medical study and practice based on reasoning from a priori assumptions about the body and illness. This approach was popular at Leiden during the second half of the seventeenth century.

As I have shown in chapters 1 and 4, he inherited his own empirical methodology and opposition to rationalist approaches from his medical mentors, Charles Drélincourt and Anton Nuck. At the same time, he incorporated into his medicine some aspects of the Cartesian approach, such as a mechanical conception of the body and its workings. He strove, however, to uphold this model of the body by grounding it in empirical observation and experimental analysis. Boerhaave has traditionally been seen as a medical mechanist, but as this book suggests, his "mechanical" medicine was, in his own understanding, also a form of experimental practice. His method of the mechanics mimicked the practitioners of mixed mathematics or "physico-mathematics," who made observations and measurements and then generated conclusions about the observed phenomena through mathematical reasoning and modeling.[2] Thus, in Boerhaave's medical system, the body was treated as a machine, but the parts and motions of that machine were to be understood empirically and experimentally. Boerhaave's version of mechanical medicine explained in part why he rejected the "Stahlian" method, which here referred to Stahl's vitalist approach to medicine. As he asserted, his students had been trained in this medical philosophy through his method of medical education, which ensured that the proper, Hippocratic approach would prevail in Leiden.

At the heart of this study is Boerhaave's rejection of "fantastical chemistry." Within the context of the medical faculty, this term most directly referred to the chemical medicine of Franciscus Sylvius, whose system dominated medical teaching at Leiden during the 1660s and 1670s. For Boerhaave he was like the Cartesians in that he based his medical system on an improper methodology. He made assumptions based on scanty evidence that the interaction of acids and alkalis controlled human health and disease. Boerhaave expressed his critique of Sylvius's approach most completely in his oration *De Chemia suos Errores Expurgante*, delivered when he became the chair of chemistry in 1718. As I discussed in chapter 4, in this oration he argued that chemists such as Robert Boyle and Johannes Bohn had rejected the excesses Sylvian chemical medicine and earlier versions of Paracelsianism after rigorously scrutinizing their claims. Chemistry was on the road to respectability, and the newly reformed and disciplined chemistry was appropriate for inclusion in the medical faculty. One of Boerhaave's aims, both in constructing his courses and inspiring his students, was to continue the work of "expunging chemistry's errors" by evaluating claims and debunking those that could not withstand experimental scrutiny. As I discussed in chapter 7, he adopted this approach outside of the lecture hall as well, most notably in his systematic, experimental critique of mercurialist alchemy published in his articles titled "De Mercurio Experimenta" (1733–36).

The core issue that framed Boerhaave's discussion with Van Royen, however, was the desire to continue attracting to Leiden students and visitors who came to study with Boerhaave. Leiden drew medical students from all over Europe and had become an essential stop for any ambitious would-be physician making his grand tour before settling into his practice.[3] By the 1730s, the number of students graduating from the medical faculty was at its highest since the 1680s, and this figure did not account for the numerous students and visitors who came to Leiden to receive Boerhaave's tutelage but did not earn a degree.[4] These men came because of Boerhaave's reputation. He was revered for the breadth of his learning in medicine and philosophy, both classical and contemporary. His "Hippocratic" medicine invoked the classical empiricist tradition of Hippocrates, which was enjoying a resurgence among medical schools, while he integrated within it modern practices in anatomy, physiology, and chemistry.[5] If Boerhaave would no longer be in the lecture hall, both he and the curators wanted to preserve and protect his methods and his legacy, because these are what brought students to Leiden.

Boerhaave opted to pass on his teaching duties to his students, because he had already handpicked them for this task. By 1721 he virtually controlled the curriculum content of the Leiden medical faculty. At this time the faculty consisted of Boerhaave; Herman Oosterdijk-Schacht (1672–1744), the town physician of Leiden and Boerhaave's opposite for the clinical practicum; and Bernard Sigfried Albinus (1697–1770), the professor of anatomy and Boerhaave's former student.[6] In addition his protégé Gerard van Swieten (1700–1772) gave private courses in medicine under his protection.[7] The ascendancy of Boerhaave's curriculum continued after he resigned the chairs of chemistry and botany in 1729. At the request of the Leiden curators, he submitted a list of possible candidates to fill these chairs and participated in the final selections. Ultimately, two more of his former students were chosen: Van Royen in botany and Hieronymus David Gaub (1705–1780) in chemistry.[8] After his resignation Boerhaave continued to offer medical courses and pursue other projects, such as the publication of *Elementa Chemiae*, the reworking of his mercury experiments, and the completion of two editing projects, the works of Aretaeus of Cappadocia and Jan Swammerdam's *Bijbel der Natuur*.[9] He fell ill in spring 1738. When the curators dispatched Van Royen to visit him, literally on his deathbed, Boerhaave suggested that, since Van Swieten could not be appointed to the medical faculty because of his Catholic religion, his remaining duties be divided among the existing professors. Six weeks before Boerhaave died on September 23, the curators charged Van Royen and Gaub with teaching the *institutiones medicae* and *praxis medica*, respectively.[10] Implicit in the curators' appointments was the understanding that Boerhaave's students would carry on their mentor's program by following his method of medical education.

Boerhaave's method was essentially his ideal curriculum, which he instituted (as far as possible) in Leiden, and his students preserved and established at other institutions. As I described in chapter 1, he crafted a vision for this curriculum in his two earliest university orations, *De Commendando Studio Hippocratico* (1701) and *De Usu Ratiocinii Mechanici in Medicina* (1703). The method received its fullest expression in a private course he offered in 1710–11, De Methodo Addiscendae Medicinae, which comprised a list of courses and the topics they should address for a comprehensive medical education. By the 1720s a version of this method was in place in Leiden as implemented by himself, his students, and other sympathetic professors such as Oosterdijk-Schacht. During the course of the eighteenth century, other medical faculties embraced the "Leiden pattern," as Gerrit Lindeboom called it.[11] Historians of medicine have long marked the spread of Boerhaave's influence throughout Europe by identifying institutions that adopted Leiden's tradition of clinical medicine as part of their curriculum.[12] As Tom Broman has shown, however, Boerhaave's textbooks were often used to ground lecture courses in physiology and pathology as well.[13] Ultimately, Boerhaave's students effected the most successful and complete transplants of the Leiden curriculum. Gerard van Swieten was called to Vienna by Empress Maria Theresa in 1745 to refurbish the medical faculty there. With the help of another Boerhaave student, Anton de Haen (1704–1776), Van Swieten reformed the curriculum, adding not only clinical instruction but also chairs in botany and chemistry. The primary teaching text in Vienna was Van Swieten's *Commentaria* on Boerhaave's *Aphorismi*, a clear nod to Boerhaave as the dominant medical authority.[14] Similarly, Albrecht von Haller (1708–1777) was named professor of anatomy, surgery, and botany at the University of Göttingen at its founding (1736), and he based his medical course on Boerhaave's *Institutiones Medicae*.[15] Four of Boerhaave's former students came to St. Petersburg, Russia, including his nephew, Abraham Kaau (1715–1758), who became the professor of anatomy at the university. When the University of Moscow was founded in 1753, each member of the medical faculty had received his medical education in Leiden.[16]

Similarly, each of the five founding members of the Edinburgh medical faculty also spent time as students in Leiden and advertised their use of Boerhaave's texts and methods in their teaching.[17] When the first medical school in America was founded in 1765 in Philadelphia, each of its professors were educated in Edinburgh, and they too followed Boerhaave's method, transplanting it to the new world.[18]

Chemistry was an integral part of Boerhaave's medical system and medical curriculum. As shown in chapter 4, he deployed chemical analysis and chemical concepts to help his students understand the composition and function of bodily substances and medicaments in his medical courses. In effect, his approach to physiology and pathology constituted a new kind of chemical medicine, which complimented the mechanistic aspects of his understanding of the body. Any medical student who wished to become fluent in Boerhaave's medical system needed to be grounded in chemistry, a fact that is reflected by chemistry's place as a foundational course in his method of medical education. As a result, medical faculties that imitated the Leiden model tended to establish chairs in chemistry or otherwise provide chemical instruction. The best examples of this practice were Edinburgh and Glasgow, where William Cullen (1710–1790) and Joseph Black (1728–1799) developed a chemistry program that included medical students and, ultimately, students of natural history and natural philosophy more generally.[19] This inclusion and integration of chemistry in medical faculties represented a significant rise in the status of chemistry among academic fields and accelerated the growth and transformation of chemistry into a scientific discipline. While at most universities the professor of chemistry was still the lowest-status member of the medical faculty, now at least chemistry had an institutional niche from which to grow.[20] The success of Boerhaave's method of medical education helped to establish chemistry as a proper university subject and, as a result, encouraged eighteenth-century medical students, physicians, and philosophers to consider the uses and problems of the field.

Unlike previous accounts of his chemistry, this study presents the *development* of Boerhaave's chemical system within its pedagogical context. I have examined several versions of Boerhaave's chemistry courses from his first lecture course in 1702 through his textbook, the *Elementa Chemiae* (1732), which combined aspects of three separate courses. During this thirty-year period, he reworked his courses, responded to new developments, and devised new methods of presentation. When he began this work, there was no standard chemistry course that fit his needs. Thus, from the morass of claims and approaches that he found in didactic courses and textbooks, chemical medicine, alchemy, and natural philosophy, he labored to establish valid facts and to derive principles from these facts to guide practice. In his own view, he worked to "expunge the errors" of chemistry and establish sound explanatory concepts and principles—the *scientia* of the chemical art. The chemistry courses that he developed through this process were guided first and foremost by traditional academic concerns about pedagogical order and structure. Yet I also contend that this work constituted a form of knowledge making: constructing a chemistry course was not the mere presentation of established facts but involved devising and developing new concepts, evaluating claims, and (when necessary) performing experiments to test them.

The biggest challenge Boerhaave faced in constructing his chemistry courses was the problem of designing them to fit the pedagogical and philosophical norms of the Leiden medical faculty. In effect, he reworked the structure and aims of the traditional didactic chemistry course so that his courses followed the same pedagogical principles and practices as his medical courses. This move allowed him to integrate chemistry into the curriculum of the medical faculty more completely than previous professors of chemistry at Leiden had done. Ultimately, he created what Peter Shaw called a "new method" for chemistry.[21] Traditional didactic courses were dominated by the presentation of recipes—descriptions of techniques to make chemical products. This approach reflected the artisanal context in which chemistry textbooks originated—as handbooks used by apothecaries and other artisan-chemists to convey practical information. By contrast Boerhaave's courses focused much more on theoretical knowledge, such as the properties of chemical species and their behavior during chemical operations. As shaped by this context, this knowledge was conveyed through academic theses that provided explanations and classifications of phenomena rather then recipes. As I argued in chapters 3 and 4, this was the type and form of knowledge that medical students had to demonstrate during their oral examinations for their medical degree and use in their medical courses proper. Boerhaave organized and presented this knowledge in a way that fit this pedagogical context and served the needs and aims of the medical curriculum.

Ultimately, Boerhaave's work represented a discrete break in pedagogical traditions of chemistry. His new method of chemistry not only reshaped the structure of the traditional chemistry course but also its content. As I have shown in chapter 3, the instrument theory, which was both the conceptual and organizational center of Boerhaave's courses, originated among academic chemist-physicians who worked on the periphery of the artisanal textbook tradition. University physicians, like Boerhaave and Johannes Bohn at Leipzig, adopted the instrument theory as a replacement for the various systems of chemical principles that predominated the typically brief discussion of chemical theory found in contemporary textbooks. Inherent in adopting the instrument theory was a change in the emphasis of chemical theory. Whereas didactic textbooks discussed the composition of chemical species in term of principles, Boerhaave, in his lectures on the instruments, examined the mechanisms through which the "natural" instruments interacted with bodies during chemical operations. Thus, the instruments—fire, air, earth, water, and chemical menstrua—functioned as the pedagogical loci around which he structured his discussion of the theoretical principles that guide chemical change and interaction. This shift in emphasis allowed Boerhaave to incorporate into his course discussion of new chemical operations and apparatus, such as air pumps, burning lenses, and thermometers. He explained how each new instrument worked through its interacting with one or more of the natural instruments and showed how they could be used to generate new chemical phenomena and new knowledge. The discussions and demonstrations performed with these instruments exemplified the new aims of Boerhaave's chemistry. They were of great importance to those interested in the theoretical principles or natural philosophy of chemistry but had little bearing on practical recipes.

Perhaps the most influential innovation in Boerhaave's pedagogy of chemistry was the new method of experimental demonstration that he first introduced in his instruments course. As I argued in chapter 5, one focus of the instruments course was to covey to students how general, theoretical claims could be generated from experimental evidence. To help him express this complicated process of manual skill, observation, and disciplined reasoning to his students, Boerhaave integrated into his course the type of demonstration experiments found in the Leiden university experimental physics and anatomical traditions. He did this in order to establish a stable understanding and protocol for new instruments, like the thermometer, that he wished to integrate into chemical practice. In the instruments course, he exhibited a series of demonstration experiments that were designed to create logical chains of inference, which connected observations with theoretical claims. For the thermometer, this chain of inference amounted to establishing the theory of how the instrument worked—the thermometric fluid expanded in proportion to the amount of instrumental fire it contained. Boerhaave, of course, was not the first chemist to present experimental work in this manner; he readily admitted that he used the work of Robert Boyle as his guide.[22] Nevertheless, the instruments course was the first course, and the *Elementa* the first textbook, to make this kind of experimental demonstration a standard part of its pedagogical method. Again, this move reflected the broader aims of Boerhaave's chemistry. For example, he wished to establish a research program for heat (fire) in which the thermometer was the central component. He outlined this program in both the instruments course and *Elementa,* ultimately encouraging his auditors and readers to obtain a Fahrenheit thermometer and continue his work.[23]

The findings presented in this volume help to establish Boerhaave's place in the historiography of eighteenth-century chemistry. We can estimate his influence by identifying those chemical practitioners who used the instrument theory and the experimental methods associated with it. In Britain, Peter Shaw adopted the instrument theory for his public chemistry lectures in London and Scarborough (ca. 1731–33), cribbing arguments and demonstration experiments right of out of the *Elementa*. In his discussion of "fire," Shaw began by examining how one can measure the degrees of fire and asserting that the expansion of solids and fluids was the only reliable measure. He then demonstrated this assertion by performing a version of Boerhaave's iron-bar experiment, showing how an iron rod, which fit through a hoop when cool, would not do so when heated. Eventually, he demonstrated that thermometers worked according to this same principle, essentially replicating Boerhaave's chain of arguments from the *Elementa Chemiae*.[24] The pedagogical and experimental work of William Cullen and Joseph Black in Edinburgh are other examples. Although both critiqued, altered, and expanded upon Boerhaave's method, the instrument theory was the center of their chemistry courses. Cullen's experiments on the generation of heat and cold and Black's work on specific and latent heats built upon the foundation of Boerhaave's thermometry and theory of fire.[25]

The instrument theory and Boerhaave's method for presenting it were also adopted and adapted for other contexts outside the university. According to Bernard Fontenelle, chemists at the Académie royale des sciences in Paris admired Boerhaave for his work on fire, which focused on chemical thermometry, and for his classification of chemical menstrua and their mechanisms.[26] French academician Pierre-Joseph Macquer (1718–1784), in his *Dictionnaire de chymie*, praised the "immortal" Boerhaave for his "admirable treatises" on air, water, earth, and "especially" fire, "which seems to leave nothing of importance for the human spirit to add."[27] At the Jardin du roi in Paris, Guillaume-François Rouelle (1703–1770) and Gabriel-François Venel (1723–1775), whose audience included apothecaries and other artisans and gentlemen interested in the practical utility of chemistry, presented in their public lectures a version of the instrument theory adapted to fit local needs and interests. As Christine Lehman has shown, both of these chemists used the instrument theory as their organizing framework, into which they integrated ideas from Boerhaave's rival, Georg Stahl, and those of other chemists. For example, they incorporated Stahl's phlogiston into Boerhaave's notion of fire: phlogiston was simply fixed fire.[28]

Boerhaave's new method could also provide practical resources for chemists. In this study, I have concentrated on the development of Boerhaave's chemistry within and for an academic context, but this was not the whole picture. As Ursula Klein has suggested, chemistry in the eighteenth century had "permeable boundaries"; artisanal and academic chemists shared knowledge and techniques as they practiced their art in diverse institutional settings.[29] Boerhaave's *Elementa* could be read as a repository for practical knowledge, and Klein argues that many chemists read the book in this way.[30] Indeed, in the *Elementa*, Boerhaave depicted his chemistry as providing knowledge and skills that were useful for various "mechanical arts"—painting, dying, glassmaking, metallurgy, wine making, brewing, and warfare—as well as for medicine and natural philosophy.[31] His main interests in this regard were in medicine and pharmacy. Combined with his *Libellus de Materie Medica* (1719), which was almost as popular as the *Elementa*, Boerhaave's textbooks provided a wealth of information on the properties and preparation of medicinal substances. The operations described in the second volume of the *Elementa* explained the medical uses of the products, when applicable.[32] This information would have been valuable to physicians, apothecaries, distillers, and other medical practitioners. Apart from practitioners of the medical arts, other artisans drew from Boerhaave's chemistry as well. In Britain, brewers began to use thermometry in their art following "Boerhaavian principles," and in Sweden, mineralogists who wished to reform their field looked to what they saw as Boerhaave's "mechanical" chemistry for concepts and methods.[33]

Ultimately, this book is one of several recent studies whose aim is to revise the historiography of early modern chemistry, especially its development leading up to the chemical revolution.[34] The traditional accounts of eighteenth-century chemistry have focused on the work of Antoine-Laurent Lavoisier, the "father of modern chemistry." This study suggests that Boerhaave and other chemists earlier in the century undertook a great deal of conceptual, experimental, and organizational work that laid the foundation for the chemical revolution. Boerhaave's *Elementa* and the chemistry courses based upon it helped to set the agenda for later chemists as they adapted his new method of chemistry for their courses and wrestled with the problems that it proposed. As Henry Guerlac has shown, Boerhaave's instrument theory shaped the general understanding of chemical activity and mechanism in Lavoisier's day, especially regarding the chemical action of fire and air.[35] Even Lavoisier acknowledged his debt to Boerhaave's work on air and thermometry. For example, he began the first chapter of his monumental *Traité élémentaire de chimie* (1789), in which he discussed the nature of heat, by stating: "That every body, whether solid or fluid, is augmented in all its dimensions by any increase of its sensible heat, was long ago fully established as a physical axiom by the celebrated Boerhaave."[36]

Boerhaave's chemistry, developed through decades of teaching, epitomized Peter Shaw's definition of "philosophical chemistry." He worked "by means of appropriate Experiments, scientifically explained, [which] lead to the discovery of *Physical Axioms*, and *Rules of Practice*, for producing useful effects" in order to "improve the State of natural knowledge, and the Arts thereon depending."[37] As I have shown in this study, Boerhaave invented a system of chemistry, which sought to generate knowledge as well as things, by establishing and organizing the precepts and principles of the art. By providing a blueprint for this new chemistry in his courses and textbook, he inspired two generations of chemists to follow this path.

Notes

INTRODUCTION

1. See Lindeboom, *Herman Boerhaave*.

2. Boerhaave's *Institutiones Medicae* (first published in 1708) went through five editions; for his medical praxis course he published *Aphorismi de Cognoscendis et Curandis Morbis* (first published in 1709) and *Libellus de Materie Medica* (1719). He published two catalogs of the Leiden botanical garden (1710 and 1720), and his chemical textbook was the *Elementa Chemiae* (1732). Boerhaave did not publish his own anatomy textbook, but he edited and published, with Leiden's anatomy professor, Bernard Albinus, an edition of Vesalius's *Opera Omnia Anatomica et Chirurgica* (1725). For a list of Boerhaave's published work, see Lindeboom, *Bibliographia Boerhaaviana*. Throughout, all translations into English are mine, except where an English-language source is indicated.

3. Underwood, *Boerhaave's Men*, 18.

4. For a list, see Kroon, "Boerhaave as Professor-Promoter."

5. See Lindeboom, *Herman Boerhaave*, 360–74; Underwood, *Boerhaave's Men*, 109–19; Emerson, "Founding of the Edinburgh Medical School"; Cunningham, "Medicine to Calm the Mind." For a graphic depiction of Boerhaave's intellectual tree, see Bohrod, "Medical Genealogy."

6. Lindeboom, *Bibliographia Boerhaaviana*, 81–86; Gibbs, "Boerhaave's Chemical Writings."

7. Lehman, "Mid-Eighteenth-Century Chemistry"; Anderson, "Boerhaave to Black."

8. Sumner, "Michael Combrune."

9. Fors, "Occult Traditions and Enlightened Science."

10. Kuhn, *Structure of Scientific Revolutions*; Kuhn, "Essential Tension."

11. Kuhn, *Structure of Scientific Revolutions*, 15; Knight, *Ideas in Chemistry*, 142–43.

12. Metzger, *Newton, Stahl, Boerhaave*. The two classic accounts of this approach in English are Thackray, *Atoms and Powers*; Schofield, *Mechanism and Materialism*. William Brock, in his recent survey of the history of chemistry, still advocates this interpretation; see Brock, *Norton History of Chemistry*, 76–77. Simon Schaffer has undermined the concept of "Newtonianism," which was the foundation of this interpretation of Boerhaave; see Schaffer, "Natural Philosophy."

13. Mauskopf, "Reflections," 185.

14. Roberts, "Going Dutch." For the British case, see Stewart, *Rise of Public Science*; Schaffer, "Natural Philosophy and Public Spectacle"; Milburn, "London Evening Courses"; Inkster, "Public Lecture."

15. See Yeo, "Organizing the Sciences," 241–66; Yeo, *Encyclopaedic Visions*; Darnton, "Philosophers Trim the Tree of Knowledge."

16. Kaiser, *Pedagogy and the Practice of Science*; Olesko, *Physics as a Calling*; Dijksterhuis, "Read-

ing Up on the *Opticks*." For recent work on chemical pedagogy, see Lundgren and Bensaude-Vincent, *Communicating Chemistry*; Bensaude-Vincent and Lehman, "Public Lectures of Chemistry"; Taylor, "Making Out a Common Disciplinary Ground."

17. Many chemists and natural philosophers read the *Elementa* in this way. See, for example, Pierre-Joseph Macquer's remarks on Boerhaave in the "Discours préliminaire" of his *Dictionnaire de chymie*, where he praised the professor's "Treatise on Fire" as "an astounding masterpiece, so complete that human understanding can scarcely make an addition to it"; Macquer, *Dictionnaire de chymie*, xxvii. See also Lehman, "Mid-Eighteenth-Century Chemistry."

18. See Meinel, "*Artibus Academicis Inserenda*," 91–92; Chang, "From Oral Disputation to Written Text."

19. Molhuysen, *Bronnen*, 1:39*–42*: "sed primus eius annis declamationi et disputationi aliquantulum, corporum vitalum, vegetabilium, et metallicorum inspectioni, dissectioni, dissolutioni ac transmutationi multum."

20. Bylebyl, "School of Padua."

21. On the founding of the University of Leiden and the medical faculty, see Otterspeer, *Het bolwerk van de vrijheid*; Lunsingh Scheurleer and Posthumus Meyjes, *Leiden University in the Seventeenth Century*; Wolter, *De Leidse Universiteit*.

22. See Lindemann, *Medicine and Society*, 172–77; Cook, "Policing Health in London." For a specific case of policing against apothecaries, see Harold Cook's analysis of the Rose case, in Cook, "Rose Case Reconsidered."

23. Hannaway, *Chemists and the Word*, 151; Moran, "Axioms, Essences and Mostly Clean Hands"; Moran, *Andreas Libavius*.

24. Christie and Golinski, "Spreading of the Word."

25. Hannaway, *Chemists and the Word*. See also Hannaway, "Laboratory Design and the Aim of Science."

26. Shackelford, "Tycho Brahe, Laboratory Design, and the Aim of Science"; Newman, "Alchemical Symbolism and Concealment."

27. Clericuzio, "Teaching Chemistry"; Joly, "Alchemie et rationalité." See also Moran, *Chemical Pharmacy*; Debus, *French Paracelsians*, 123–46.

28. Newman and Principe, "Alchemy vs. Chemistry."

29. Principe, "Revolution Nobody Noticed?" See also Powers, "'*Ars sine arte*.'"

30. See Hotson, *Commonplace Learning*, 44; Ong, *Ramus*, 112–16, 182–90. The title of this book was inspired by the title of chap. 6 ("Chemistry Invented") in Hannaway, *Chemists and the Word*.

31. On classical views of "art," see Kristeller, "Modern System of the Arts"; Gilbert, *Renaissance Concepts of Method*, 11–13, 69–71; Hannaway, *Chemists and the Word*, 127–29; Ong, *Ramus*. On pedagogical method in the sciences, see Dear, "Method and the Study of Nature."

32. Yeo, "Organizing the Sciences," 241. For an example of the search for *scientia*, see Henninger-Voss, "How the 'New Science' of Cannons."

33. Donovan, *Philosophical Chemistry*. See also Christie, "Cullen and the Practice of Chemistry"; Taylor, "Unification Achieved."

34. Shaw, *Chemical Lectures*, 1.

35. Golinski, "Utility and Audience"; Golinski, *Science as Public Culture*; Stewart, *Rise of Public Science*; Briggs, "Académie Royale des Sciences."

36. On the "continuity thesis," see Newman and Principe, "Alchemy vs. Chemistry"; Principe, "Revolution Nobody Noticed?"; Principe, "Wilhelm Homberg"; Powers, "'*Ars sine arte*.'" The founder of this approach is Allen Debus; see Debus, "Alchemy in an Age of Reason"; Debus, *French Paracelsians*, chaps. 5 and 6.

37. A published catalog of this archive with commentary may be found in Cohen and Cohen-de Meester, "Katalog der Wiedergefunden." The archive has been reordered since the publication of Cohen and Cohen-de Meester's catalog. The proper lot number for each item may be found in a

second catalog without commentary: "Lijst van de Boerhaaviana in de S. M. Kirow-Academie te Leningrad," in Schulte, *Hermanni Boerhaave Praelectiones de Morbis Nervorum*, 426–33.

38. See Cook, *Matters of Exchange*; Cook, "New Philosophy in the Low Countries"; Ruestow, *Microscope in the Dutch Republic*; Alpers, *Art of Describing*.

39. Knoeff, *Herman Boerhaave*; Cook, "Boerhaave and the Flight from Reason in Medicine"; Cook, *Matters of Exchange*, 383–409.

CHAPTER ONE: MEDICINE AS A CALLING

1. As quoted in Atkins, "Johnson's 'Life of Boerhaave,'" 106.

2. Von Haller, *Albrecht Hallers Tagebücher*, 31, 98.

3. Burton, *Account of Boerhaave*, 61–62.

4. Weber, *Protestant Ethic*, 98–128. Although Weber's main thesis, the connection between Calvinism and capitalism, is now questioned, Weber's characterization of the Calvinist character, especially by the time of the eighteenth century, is valid; see, for example, Green, *Protestantism, Capitalism, and Social Science*; Lehmann and Roth, *Weber's "Protestant Ethic"*; Schama, *Embarrassment of Riches*, esp. 323–43. See, especially, regarding Protestant self-fashioning, Goldman, "Weber's Ascetic Practices."

5. Schultens, *Oratio Academica*, 72. This anecdote was also mentioned by Samuel Johnson; see Atkins, "Johnson's 'Life of Boerhaave,'" 106.

6. Cunningham, "Medicine to Calm the Mind," 48. On "eirenicism" in English medicine, see Elmer, "Medicine, Religion, and the Puritan Revolution," esp. 34–45.

7. Cook, "Boerhaave"; Cook, "Body and Passions." See also Dear, "Mechanical Microcosm"; Cunningham, "Medicine to Calm the Mind," 46–56.

8. Knoeff, *Herman Boerhaave*; Knoeff, "Making of a Calvinist Chemist."

9. Lindeboom, *Herman Boerhaave*, 247–49.

10. On Bacon's methodology, see Dear, "Method and the Study of Nature"; Farrington, *Philosophy of Francis Bacon*.

11. On "physico-mathematics," see Dear, *Discipline and Experience*.

12. Cunningham, "Medicine to Calm the Mind"; Cook, "Boerhaave"; Beukers, "Clinical Teaching"; Frijhoff, *Société néederlandaise*, 103–7.

13. Boerhaave, "Commentariolus," 375–86.

14. See Schultens, *Oratio Academica*; Burton, *Account of Boerhaave*. On other eighteenth-century biographies of Boerhaave, see Lindeboom, *Herman Boerhaave*, 238–40.

15. See Lindeboom, *Herman Boerhaave*, 11–12, 14.

16. Boerhaave, "Commentariolus," 377: "Vir apertus, candidus, simplex: Paterfamilias optimus amore, cura, diligenta, frugalitate, prudentia. Qui non magna in re, sed plenus virtutis, novem liberis educandis exemplum praebuir singulare, quid exacta parsimonia polleat, & frugalitas."

17. Boerhaave, "Academic Discourse, Delivered by Herman Boerhaave When He Officially Resigned His Professorships in Botany and Chemistry, Having Obtained an Honourable Discharge, on 28 April 1729," in *Boerhaave's Orations*, 214–36, at 225. On Dutch permissiveness toward children, see Zumthor, *Daily Life*, 99–102; Schama, *Embarrassment of Riches*, 485–86.

18. See, for example, Schama's discussion of the ideology behind the "workhouses" established for delinquent Dutch youth, in Schama, *Embarrassment of Riches*, 15–24.

19. On Latin schools in the Dutch Republic and their curriculum, see Zumthor, *Daily Life*, 110–12; Cook, *Trials of an Ordinary Doctor*, 47–48.

20. Boerhaave, "Commentariolus," 378. See also Lindeboom, *Herman Boerhaave*, 15–18.

21. Schama, *Embarrassment of Riches*, 404–5.

22. Resoluties van Curatoren (Res. Cur.), 3 December 1687, in Molhuysen, *Bronnen*, 4:54.

23. Ruestow, *Physics*, 45–46, 61–72; Israel, *Dutch Republic*, 890, 891; McGahagan, "Cartesianism in the Netherlands," 231–61; Verbeek, *Descartes and the Dutch;* Vermij, *Calvinist Copernicans*, 137–237.

24. Res. Cur., 11 January 1676, in Molhuysen, *Bronnen*, 3:319: "tot heter hanthavinge van de goede saeken." See also De Pater, "Experimental Physics," 319; Van Poelgeest, "Stadholder-King," 111–13.

25. Senguerdius, *Philosophia Naturalis*, 36, 80, 160–86, 245–71; Anon., "*Philosophia Naturalis.*" See also De Pater, "Experimental Physics," 319–20; Ruestow, *Physics*, 78–87.

26. Lindeboom, *Herman Boerhaave*, 21–22; Lindeboom, *Bibliographia Boerhaaviana*, nos. 1–4. The title of the fifth disputation is lost. For the university disputation tradition, see Chang, "From Oral Disputation to Written Text"; Grant, *Foundations of Modern Science*, 39–42; Sylla, "Science for Undergraduates."

27. Lindeboom, *Herman Boerhaave*, 22–23. For the establishment of the award and Boerhaave's reception of it, see, respectively, Res. Cur., 3 December 1687 and 15 November 1690, in Molhuysen, *Bronnen*, 4:54, 86. The curators designated that the medal should be worth about fifty guilders.

28. Boerhaave, "Oration on the Thesis That Cicero's Interpretation of Epicurus' Maxim on the Highest Good Is Right," in *Boerhaave's Orations*, 18–53; Sassen, "Intellectual Climate," 4–5; Cook, *Matters of Exchange*, 384–85.

29. Several historians have suggested that Boerhaave's student oration publicly rejected Senguerd's "Gassendist" natural philosophy as a way to distance himself from his old mentor. I do not think Boerhaave intended this, because later he incorporated aspects of Senguerd's approach in his own work, especially regarding the concept of "fire" as a subtle, particulate fluid; see chap. 3. On this interpretation of the oration, see Cook, *Matters of Exchange*, 385; Sassen, "Intellectual Climate," 5; Kegel-Brinksgreve and Luyendijk-Elshout, *Boerhaave's Orations*, 20.

30. "Dictata viri clariss. Bucherii de Volder in principia cartesii," in MS 1216, British Library, Sloane Collection, 75r–128v; Ruestow, *Physics*, 92–95; Sassen, "Intellectual Climate," 6–7. The best summary of De Volder's life and beliefs, if biased, remains LeClerc, "Eloge."

31. De Pater, "Experimental Physics," 315–17; LeClerc, "Eloge," 398; Ruestow, *Physics*, 94–104; Wiesenfeldt, *Leerer Raum*, 108–88. Student notes of De Volder's 1676 course may be found at the British Library, Sloane Collection, MS 1292. For an inventory of instruments found in the *theatrum* in 1705 (at De Volder's retirement), see "Lijste en register van de machininen en instrumenten behoorende tet Theatrum Experimentale Physicum," 14 November 1705, in Molhuysen, *Bronnen*, 4:104*–6*. On Musschenbroek and his instruments, see de Clercq, *Sign of the Oriental Lamp*.

32. Ruestow, *Physics*, 105–8; Klever, "Burchard de Volder," 207–10. On "physico-mathematics," see Dear, *Discipline and Experience*.

33. Ruestow, *Physics*, 110–11.

34. Boerhaave, "Commentariolus," 379: "quae mirifice ingenio ejus placebat. Synthesin geometricam veterum admirans maxime, & excolens, ad augendam vim Intelligentiae; Analysin Recentiorum ad usum nova inveniendi."

35. Boerhaave, *De Distinctione*. See Wright, "Boerhaave," 290.

36. Burton, *Account of Boerhaave*, 13–15; Lindeboom, *Herman Boerhaave*, 25–27; Kegel-Brinkgreve and Luyendijk-Elshout, *Boerhaave's Orations*, 54–55. On Boerhaave's appointment as assistant to the "keeper," see Res. Cur., 18 October 1691 and 3 November 1692, in Molhuysen, *Bronnen*, 4:96, 107. On the Vossian collection, see Rademaker, "Famous Library."

37. Claims regarding Boerhaave's autodidactic training in medicine may be found in, for example, Burton, *Account of Boerhaave*, 14–16; Fontenelle, "Eloge de M. Boerhaave"; and Samuel Johnson's "Life of Boerhaave."

38. Boerhaave, "Commentariolis," 381; Lindeboom, *Herman Boerhaave*, 30–31, 36; Lindeboom, "Frog and Dog," 290–91.

39. Boerhaave, "Commentariolus," 381; Lindeboom, *Herman Boerhaave*, 30–32.

40. Boerhaave, "Commentariolis," 379–80; Burton, *Account of Boerhaave*, 10–11.

41. Boerhaave, "Commentariolis," 380: "Horum simplicitatem sincerae doctrinae, disciplinae sanctitatem, vitae DEO dicate integritatem adorabat; subtilitatem scholarum Divina postmodum inquinasse dolebat."

42. Cunningham, "Medicine to Calm the Mind," 45–46.

43. On the debate in Leiden over the moderate theology of Johannes Cocceius (1603–1669), see Israel, *Dutch Republic*, 662–69, 838–40, 897–99.

44. Boerhaave, "Commentariolus," 381; Burton, *Account of Boerhaave*, 14–15. Quotation from Boerhaave, "Commentariolus," 381: "cito deprehendit posteriores omnia bona sua Hippocrati debere." On Sydenham, see Cunningham, "Thomas Sydenham."

45. Cunningham, "Medicine to Calm the Mind," 46–56.

46. Boerhaave, in fact, offered a course (1703) on the various medical sects in which he explained how they differed and, ultimately, fell into error. See "De Sectis Medicorum Praelectiones," VMA, MS 19.

47. Penning, "De Promotie"; Lindeboom, *Herman Boerhaave*, 38–42.

48. Burton, *Account of Boerhaave*, 19–21.

49. Israel, *Dutch Republic*, 896–97, 916–25. On Spinoza's views and their place in Dutch culture, see Israel, *Radical Enlightenment*, 157–326; Van der Wall, "*Tractatus Theologico-Politicus*."

50. Boerhaave, "Commentariolus," 382; Lindeboom, *Herman Boerhaave*, 45–47. The best examination and analysis of this episode is Knoeff, *Herman Boerhaave*, 30–46.

51. Boerhaave, "Commentariolus," 382.

52. Lindeboom, *Herman Boerhaave*, 264–66; Boerhaave to Cox Macro, 22 June 1710, in *Boerhaave's Correspondence*, 1:18–21.

53. Klever, "Burchard de Volder."

54. Knoeff, *Herman Boerhaave*, 43–46.

55. Boerhaave to Cox Macro, 22 June 1710.

56. See Boerhaave, "Commentariolus," 380.

57. Ibid., 382–83; Burton, *Account of Boerhaave*, 21–23. Quotation from Boerhaave, "Commentariolus," 382: "Contentus videlicet vita libera, remota a turbis, studiisque porro percolendis unice impensa, ubi non cogeretur alia dicere & simulare, alia sentire & dissimulare: affectuum studiis rapi, regi."

58. The university's debt amounted to 20,000 guilders by 1696, about one-third of its annual budget. This debt was paid with a grant from the States of Holland in 1699. See Van Poelgeest, "Stadholder-King," 130–31; "Extract uit de Resoluties van de Staten van Holland betr. den aankoop van de Bibliotheca Vossiana," 28 July 1690, and "Extract uit de Resoluties van de Staten van Holland," 17 January 1699, in Molhuysen, *Bronnen*, 4:25*–26* and 68*–70*.

59. William, in fact, styled himself the "Opper-curator" of the university. He used his authority to intervene in university affairs and make political appointments on the curators and faculty. See Van Poelgeest, "Stadholder-King," 107–21. On Le Mort, see Molhuysen, *Bronnen*, 4:161–67; Lindeboom, "Mort, Jacobus le," in *Dutch Medical Biography*, cols. 1369–71; Wiesenfeldt, *Leerer Raum*, 207–10, 219–13.

60. Ultee, "Politics of Professorial Appointment," 171; Lindeboom, *Herman Boerhaave*, 51–52. On Hotton, see Lindeboom, "Hotton, Petrus," in *Dutch Medical Biography*, cols. 912–13. On Dekkers, see Lindeboom, "Deckers (Dekkers), Frederik," in *Dutch Medical Biography*, cols. 412–13.

61. On medical faculty degrees, see Frijhoff, *Société néederlandaise*, 193–95 (esp. fig. 3).

62. Res. Cur., 18 May 1701, in Molhuysen, *Bronnen*, 4:189.

63. Res. Cur., 4 January 1702, ibid., 4:190: "op de 'ernstige instantien van eenige vremde studenten.'"

64. Le Mort was appointed chair of chemistry; Govert Bidloo (1649–1713), who had been one of William's personal physicians, returned to Leiden to resume his duties as chair of anatomy; Bernard Albinus (1653–1721) was appointed chair of the Institutiones Medicae. See Lindeboom,

Herman Boerhaave, 51–52; Israel, *Dutch Republic,* 959–62; Res. Cur., 6 May 1702, in Molhuysen, *Bronnen,* 4:197; "Verzoek van C. en B aan de Staten om subsidie," 14 July 1702, in Molhuysen, *Bronnen,* 4:92*–93*.

65. He was also granted a 200-guilder per annum raise and permission to give an academic oration to the university community. Res. Cur., 12 April and 8 May 1703, in Molhuysen, *Bronnen,* 4:208. See also Burton, *Account of Boerhaave,* 28–29.

66. Res. Cur., 18 February 1709, in Molhuysen, *Bronnen,* 4:242. On the minor controversy over Boerhaave's appointment, see Ultee, "Politics of Professorial Appointment."

67. Burton, *Account of Boerhaave,* 61–62.

68. Boerhaave, *Institutiones Medicae;* Boerhaave, *Aphorismi.* See also Lindeboom, *Herman Boerhaave,* 70–78.

69. "Collegium Chemicum," VMA, MS 3, 1r, 143r; "De Sectis Medicorum Praelectiones," VMA, MS 19; Lindeboom, *Herman Boerhaave,* 59, 68.

70. Boerhaave, *Oratio de Commendando Studio Hippocratico.* I quote from the English translation in Boerhaave, "Oration to Recommend the Study of Hippocrates," in *Boerhaave's Orations,* 54–84, at 79, 83.

71. On the Hippocratic revival in medicine, see Nutton, "Hippocrates in the Renaissance"; Lonie, "'Paris Hippocrates'"; Cantor, *Reinventing Hippocrates.*

72. Boerhaave, "Oration to Recommend the Study of Hippocrates," in *Boerhaave's Orations,* 65–66, 68, 69.

73. On this count, Boerhaave stated: "He [Hippocrates] is entitled to the claim . . . that he never invented what he had not actually observed; that he never twisted or meddled the truth when describing the works of nature, so as to achieve lasting fame for an hypothesis that would otherwise have been seen to be shaky." Ibid., 70.

74. Ibid., 74–75.

75. Ibid., 81–82.

76. "Forma Examinis, qua Doctor Medicinae," in Molhuysen, *Bronnen,* 1:48*; Lindeboom, "Medical Education," 203.

77. See Boerhaave's praise of Sydenham's "Hippocratic method," in Boerhaave, "Oration to Recommend the Study of Hippocrates," in *Boerhaave's Orations,* 78, 83; Boerhaave, "Commentariolus," 381. On Hippocrates in Sydenham's work, see Cunningham, "Thomas Sydenham." Although Boerhaave did not cite Bacon in his *Oratio de Commendando,* the work was very Baconian in character. Boerhaave's first public praise of Bacon was in his 1715 oration *Sermo Academicus de Comparando Certo in Physicis.* Bacon's description of Hippocrates's method was very similar to Boerhaave's; see Bacon, *Advancement of Learning,* 112–13. On the popularity of Bacon's work in the Dutch Republic, see Elena, "Baconianism in the Seventeenth-Century Netherlands."

78. Boerhaave, "Oration to Recommend the Study of Hippocrates," in *Boerhaave's Orations,* 72–73, 76–78.

79. Ibid., 72, 73, 75, 77. Debate on the "authentic" Hippocratic writings began in antiquity and continues unresolved to this day. As G. E. R. Lloyd pointed out, historically most physicians and scholars who have attempted to identify the authentic corpus did so by reading their own preconceived notions into the project; see Lloyd, "Hippocratic Question."

80. Boerhaave, "Oration to Recommend the Study of Hippocrates," in *Boerhaave's Orations,* 75.

81. Ibid., 80.

82. On "constructive geometry," see Dear, *Discipline and Experience,* 210–27. Regarding this idea among mathematical practitioners and natural philosophers, see Garrison, "Newton"; Bennet, "Mechanics' Philosophy"; Bennet, "Robert Hooke."

83. Boerhaave, "Oration to Recommend the Study of Hippocrates," in *Boerhaave's Orations,* 80.

84. See Boerhaave's second oration, *De Usu Ratiocinii Mechanici in Medicina.* I quote from the English translation in Boerhaave, "On the Usefulness of the Mechanical Method in Medicine," in *Boerhaave's Orations,* 116–17. See also "Oration to Recommend the Study of Hippocrates," 81, 83.

85. Boerhaave, "Oration to Recommend the Study of Hippocrates," in *Boerhaave's Orations*, 74.

86. Ibid., 82.

87. Schama, *Embarrassment of Riches*, 323–43.

88. Boerhaave, "Oration to Recommend the Study of Hippocrates," in *Boerhaave's Orations*, 75.

89. Boerhaave, "On the Usefulness of the Mechanical Method in Medicine," in *Boerhaave's Orations*, 99, 103, 109.

90. Boerhaave, *Oratio qua Repurgatae Medicinae Facilis Asseritur Simplicitas.*

CHAPTER TWO: DIDACTIC CHEMISTRY IN LEIDEN

1. On "mercurialist" alchemy, see Newman, *Gehennical Fire*; Principe, "Diversity in Alchemy."

2. Newman, "Authorship"; Newman, *Gehennical Fire.*

3. See Newman, *Gehennical Fire*, chap. 4.

4. VMA, MS 131, 86r–v.

5. This operation was described by Lemery, *Cours de chymie*, 296–303. The ability of crude antimony to "absorb" base metals was well known by assayers; see Ercker, *Fleta Minor*, chaps. 49–51; Agricola, *De Re Metallica*, 237–39, 451–52.

6. Newman, "Authorship," 139–44; Newman, *Gehennical Fire.*

7. See Newman, *Gehennical Fire*, 53–91.

8. Hannaway, *Chemists and the Word.*

9. Christie and Golinski, "Spreading of the Word."

10. Joly, "Alchemie et rationalité"; Clericuzio, "Teaching Chemistry"; Newman and Principe, "Alchemy vs. Chemistry." See also Klein, "Nature and Art"; Moran, *Distilling Knowledge*, chaps. 3–4; Debus, *French Paracelsians*, 123–34.

11. Hannaway, *Chemists and the Word*; Moran, *Andreas Libavius.*

12. Moran, *Andreas Libavius*, 1–10, 31–49; Hannaway, *Chemists and the Word*, 122–23.

13. Hannaway, *Chemists and the Word*, 117–51. See also, Moran, "Axioms, Essences and Mostly Clean Hands."

14. Libavius, *Alchemia*, 1; Hannaway, *Chemists and the Word*, 143.

15. On Ramus's method and its pedagogical context, see Ong, *Ramus*; Gilbert, *Renaissance Concepts of Method*; Hannaway, *Chemists and the Word*, 124–41. On the spread and influence of Ramism in Germany, see Hotson, *Commonplace Learning.*

16. Libavius, *Alchemia*, 1.

17. Hannaway, *Chemists and the Word*, 142–51.

18. Ibid., 151.

19. On seventeenth-century textbooks, see Hannaway, *Chemists and the Word*, 152–56; Clericuzio, "Teaching Chemistry"; Klein, "Nature and Art"; Thorndike, *History of Magic*, 8:104–69; Metzger, *Doctrines chimiques.*

20. Clericuzio, "Teaching Chemistry," 339–41. On the Paracelsian controversy in France, see Debus, *French Paracelsians*, 17–101.

21. See Kent and Hannaway, "Some New Considerations."

22. Patterson, "Jean Beguin"; Beguin, *Tyrocinium.*

23. Patterson, "Jean Beguin," 250–51.

24. Thorndike, *History of Magic*, 8:104–69.

25. Hannaway, "Lemery, Nicolas." On Lemery, see Clericuzio, "Teaching Chemistry," 347–50; Holmes, "Investigation and Pedagogical Style"; Bougard, *Chimie du Nicolas Lemery.*

26. Clericuzio, "Teaching Chemistry," 343–47; Debus, *French Paracelsians*, 80–84, 123–41; Howard, "Guy de la Brosse"; Guerlac, "Guy de la Brosse."

27. See Klein, "Apothecary-Chemists." On craft training generally, see Smith, "Sixteenth-Century Goldsmith's Workshop"; Long, *Openness, Secrecy, Authorship.*

28. See Susan Lawrence's notion of "educating the senses" in eighteenth-century anatomy courses as a useful comparison; Lawrence, "Educating the Senses."

29. Moran, *Chemical Pharmacy*, 45–51.

30. Fontenelle, "Eloge de Nicolas Lemery," 96–108. See also Sutton, *Science for a Polite Society*.

31. Beguin, *Tyrocinium*, 1.

32. Patterson, "Jean Beguin," 252–57; Beguin, *Tyrocinium*, 2–27.

33. Beguin, *Tyrocinium*, 40. For the *Summa Perfectionis*, see Newman, *Summa Perfectionis*, 417 (Latin), 704 (English). The margin note in Beguin indicates that this passage was found in chap. 52 of the *Summa*, but the passage appears in chap. 47 of Newman's critical edition.

34. Beguin, *Tyrocinium*, 97–99.

35. Ibid., 120–42.

36. Lemery, *Cours de chymie*.

37. On Hartmann, see Moran, *Chemical Pharmacy*, 337–39.

38. Turnbull, "Peter Stahl." For subsequent developments, see Guerrini, "Chemistry Teaching."

39. Van Spronsen, "Beginning of Chemistry."

40. On early interest in chemistry at Leiden, see ibid., 333. For the study of chemistry at apothecary's shops, see the cases of Robert Sibbald (ca. 1660–61), described in Cook, *Trials of an Ordinary Doctor*, 63; and Herman Boerhaave (ca. 1691–93) in Lindeboom, *Herman Boerhaave*, 35; Burton, *Account of Boerhaave*, 16.

41. "Rekest van den apotheker Chimaer aan C. en B. om salaris te ontvangen voor zijn onderwijs in de pharmacognosie," 8 February 1642, in Molhuysen, *Bronnen*, 2:399*–400*. On the *collegium medico-practicum*, see Beukers, "Clinical Teaching," 140–44; "Advies van die Medische Faculteit over het invoeren van klinisch onderwijs," 20 September 1636, in Molhuysen, *Bronnen*, 2:314*–15*.

42. On Sylvius, see Baumann, *François Dele Boë Sylvius*; King, *Road to Medical Enlightenment*, 93–113; Foster, *Lectures*, 147–62; Partington, *History of Chemistry*, 2:281–90; Underwood, "Franciscus Sylvius," 73–76; Beukers, "Acid Spirits and Alkaline Salts."

43. For example, the Scottish student Robert Sibbald reported that in 1660–61 he observed twenty-three dissections at the St. Catherine and Cecilia Hospital under Sylvius's direction. Sylvius's course (the clinical practicum) was so popular that the dissection theater had to be enlarged. See Cook, *Trials of an Ordinary Doctor*, 63; Lindeboom, "Medical Education," 206; Res. Cur., 8 February 1660, in Molhuysen, *Bronnen*, 3:158.

44. Res. Cur., 8 May 1666, in Molhuysen, *Bronnen*, 3:204. The hospital was a charitable institution funded by the town to serve the poor of Leiden, so any medicaments were paid for by the town. The curators and burgemeesters agreed to pay 120 guilders per year to cover the additional costs of Sylvius's prescriptions.

45. British Library, Sloane Collection, MS 1287. Three versions of Schacht's courses are included in this manuscript (1r–109v, 110r–62v, and 163r–236v), along with a course on pharmacy by Jacob le Mort (236r–77v). The Le Mort course is dated 1678, but unfortunately the Schacht courses are undated.

46. On Deusing, see Hoving-Ebels, "Theorie en therapie," 49–68; Vermij, *Calvinist Copernicans*, 121–23. A pamphlet that Sylvius wrote against Deusing, *Epistola Apologetica Improbas aequae ac Ineptas Antonii Deusingii*, may be found in Sylvius, *Opera Medica*, 905–11.

47. Res. Cur., 8 February 1666, in Molhuysen, *Bronnen*, 3:203–4; "Acte van redenheid voor Prof. Sylvius," 8 February 1666, ibid., 196*; Baumann, *François Dele Boë Sylvius*, 33–35, 67–69; Van Spronsen, "Beginning of Chemistry," 336. On Sylvius's private laboratory, see Beukers, "Het Laboratorium van Sylvius."

48. The rivalry between Utrecht (founded in 1636) and Leiden for students and status was ongoing during the seventeenth century and had a significant impact on the development of both schools. For example, the Leiden medical faculty initiated its medical practicum at the St. Catherine and Cecilia Hospital largely because a similar program was being discussed at Utrecht. See Lindeboom, "Medical Education," 204–6; Beukers, "Clinical Teaching," 140.

49. Res. Cur., 8 February 1669, in Molhuysen, *Bronnen*, 3:227–28. On De Maets, see Van Spronsen, "Beginning of Chemistry," 336–38; Lindeboom, "Maets (Dematius), Carel," in *Dutch Medical Biography*, cols. 260–61.

50. Res. Cur., 12 January 1671, in Molhuysen, *Bronnen*, 3:247; Van Spronsen, "Beginning of Chemistry," 337, 341.

51. Lindeboom, "Maets (Dematius), Carel"; Res. Cur., 15 August 1672, in Molhuysen, *Bronnen*, 3:262.

52. Res. Cur., 3 January 1679, in Molhuysen, *Bronnen*, 3:341.

53. This term was applied to various uneducated medical practitioners, including chemists and apothecaries; see Wear, "Medical Practice"; Bynum and Porter, *Medical Fringe*. For chemists as empirics, see Principe, *Aspiring Adept*, 30–35.

54. Res. Cur., 3 December 1674 and 26 January 1675, in Molhuysen, *Bronnen*, 3:298, 301–2, respectively. On De Volder's experimental physics courses and the Leiden physics theater, see Wiesenfeldt, *Leerer Raum*, 99–132; De Pater, "Experimental Physics," 309–27.

55. "Collegium Chemicum Clarissimi, Expertissimque Viri D. Carlosi de Maets," in VMA, MS 131, 56r: "Chemiae recte considerata. . . . ut pote_ p distillationes, sublimationes, digestiones, putrefactiones &c: . . . numerat medicinas, partemque ipsi[am] [*sic*] pharmaceuticam [con]stituit & indagat Pharmacopoea Spagyrica, hermetica, distillatonia."

56. Ibid., 56r (introduction), 56v–95r (operations).

57. British Library, Sloane Collection, MS 1286, 57v–58v.

58. VMA, MS 131, 63r, 83r–v. On early chemical indicators, see Eamon, "New Light."

59. British Library, Sloane Collection, MS 1286, 93r–v.

60. Ibid., 94r–96v. On the mercury/sulfur dyad, see Principe, *Aspiring Adept*, 36–42; Newman, *Summa Perfectionis*; Newman, *Gehennical Fire*, 96–97.

61. Lindeboom, "Marggraff (Markgraave, Margravius, Merckgravius), Christiaan," in *Dutch Medical Biography*, cols. 1267–69; Van Spronsen, "Beginning of Chemistry," 340–41. Marggraf's petition of the academic senate was recorded in Acta Sen., 6 March 1659, in Molhuysen, *Bronnen*, 3:142.

62. "Collegium Chymicum Doctissimi Viri Dri. Cristiani Marggravii," in VMA, MS 131, 105r–49r; British Library, Sloane Collection, MS 1284. This manuscript (Sloane 1284) contains four versions of Marggraf's course, although not all are complete.

63. Van Spronsen, "Beginning of Chemistry," 339, 441; Lindeboom, "Mort, Jacobus le (Le)," in *Dutch Medical Biography*, cols. 1369–71; Partington, *History of Chemistry*, 2:737; Wiesenfeldt, *Leerer Raum*, 207–22.

64. Le Mort, *Compendium Chymicum*, 2–11; British Library, Sloane Collection, MS 1286. Three versions of Le Mort's course are contained here (112r–end), along with two by De Maets.

65. British Library, Sloane Collection, MS 1287; Le Mort, *Pharmacia Medico-Physico*.

66. British Library, Sloane Collection, MS 1286, 70r–111v ("*Excerpta Maetsiana*"); MS 1284, 1r–62v ("*Excerpta Chymica . . . Maargraviani*").

67. Morley, *Collectanea Chymica*. See Van Spronsen, "Beginning of Chemistry," 338; Wiesenfeldt, *Leerer Raum*, 202–6. The book was most likely printed in London, where Henry Drummond worked as a publisher, not in Leiden. Morley's original course notes are held at the British Library, Sloan Collection, MSS 1284–86.

68. Wiesenfeldt, *Leerer Raum*, 204n44.

69. Van Spronsen, "Beginning of Chemistry," 340; Wiesenfeldt, *Leerer Raum*, 206. See, for example, the pamphlets against each other that Marggraf and Le Mort published: Marggraf, *Jacobi le Mort, pseudochemici*; Le Mort, *Ignorantia circa Chemiam*.

70. De Maets, *Prodromus*, 4: "Furta haec sunt & crimina viris doctis intoleranda."

71. Ibid., 7–11.

72. See Powers, "'*Ars sine arte*'"; Principe, "Revolution Nobody Noticed."

73. De Maets, *Prodromus*, 6: "Nonne hic crassus error, & hoc temeritatis tuae argumentum est?"

For the printer's original remarks, see Morley, *Collectanea Chymica,* fifth page of unpaginated "Typographicus Lectori."

74. De Maets, *Prodromus,* 6, 13–14, quotation at 13: "Non eniam excogitandum neque figendum (inquit magnus ille VERLAMIUS) sed inveniendum quid Natura faciat aut ferat."

75. Ibid., 14–15: "fingendo Hypotheses dubias, obscuras, immo falsas, in mille incideret absurda, neutiquam cum structura machinae nostrae, motubusque quibus regitur convenientia?"

76. "Advies van de Medische faculteit over privaat-colleges van niet-professoren," 22 March 1690, in Molhuysen, *Bronnen,* 6:23*. The memorandum named Le Mort and a Cartesian medical lecturer, Johannes Broen.

77. Res. Cur., 8 November 1690, in Molhuysen, *Bronnen,* 6:84–85.

78. Le Mort, *Chymia Verae Nobilitas.*

79. On William of Orange's intervention in university affairs, see Van Poelgeest, "Stadholder-King."

80. Lindeboom, *Herman Boerhaave,* 109–10.

81. Boerhaave, "*Commentariolus,*" 381.

82. VMA, MS 131.

83. Boerhaave, *Elements,* unpaginated "Dedication."

84. On Stam, see Lindeboom, "David en Nicolaas Stam."

85. Burton, *Account of Boerhaave,* 16.

86. Vigani, *Medulla Chymiae.* On Vigani, see Coleby, "John Francis Vigani"; Guerrini, "Chemistry Teaching," 186–88; Partington, *History of Chemistry,* 2:686–87.

87. Vigani, *Medulla Chymiae,* 11–13.

88. Stam, "Proaemium" in Vigani, *Medulla Chymiae,* 1–9; at 1: "hoc Naturae vocabulum paulo clarius illustrem."

89. Stam, footnote, in Vigani, *Medulla Chymiae,* 13–18. Van Helmont's position on the chymical principles was much more complex and ambivalent than that presented by Stam. See Newman, *Gehennical Fire,* 110–14.

90. Boerhaave, *Elements,* 1:52, 56.

91. Knoeff, *Herman Boerhaave,* 181.

92. On the experimental connection between Van Helmont's and Boyle's research, see Newman and Principe, *Alchemy Tried in the Fire.*

93. Most of these notes are contained in two bound volumes: VMA, MSS 6 and 137.

94. VMA, MS 1, 32r–33v. Boerhaave also records passages from the work of Otto Tachenius and Johannes Bohn here.

95. VMA, MS 6, 110r–11r. The passages in Boerhaave's notes are taken from "Progymnasma meteroi," 70–71, and "Tria Prima Chymicorum Principia," 407–8, in Van Helmont, *Ortus Medicinae.*

96. On the *alkahest,* see Newman, *Gehennical Fire,* 146–48 and passim; Porta, "'*Summus atque felicissimus salium*'"; Joly, "L'alkahest."

97. Principe, *Aspiring Adept,* 37–39. On the origins of this theory, see Norris, "Mineral Exhalation Theory."

98. Newman, *Summa Perfectionis;* Newman, "Technology and Alchemical Debate."

99. VMA, MS 137. The notes on the *Summa Perfectionis* are found at 88r–93v.

100. VMA, MS 1, 85r. Philalethes's version of this process was found in Philalethes, *Introitus.* The "eagles" are mentioned on 685. For interpretation of Philalethes's text, see Newman, *Gehennical Fire,* 115–69.

101. VMA, MS 1, 85r–86v.

102. See Ercker, *Fleta Minor,* chaps. 49–51; Agricola, *De Re Metallica,* 237–39, 451–52.

103. For the mercurialist interpretation of this operation, see Newman, *Gehennical Fire,* 128–33; Newman and Principe, *Alchemy Tried in the Fire.* Boerhaave later interprets this operation in a similar way; see Boerhaave, *Elements,* 2:358–60.

104. See Debus, *French Paracelsians*, 74–80, 95–99; Multhauf, *Origins of Chemistry*, 201–36.

105. See Newman, *Gehennical Fire*; Newman and Principe, *Alchemy Tried in the Fire*.

106. Lemery, *Traité de l'antimoine*. The experimental work for this study was completed from 1699 through 1705.

107. On sulfur and antimony in Van Helmont, see Newman and Principe, *Alchemy Tried in the Fire*, 101–14. On Sala, see Gelman, "Angelo Sala"; Thorndike, *History of Magic*, 7:167–70.

108. VMA, MS 1, 27r–28r.

109. Ibid., 17r–18r, 21r. On metallic preparations as medicines, see Debus, *French Paracelsians*, 74–80, 95–99; Multhauf, *Origins of Chemistry*, 201–36.

110. See Newman and Principe, *Alchemy Tried in the Fire*, 114–17.

111. VMA, MS 1, 18r.

112. Boerhaave, *Elements*, 1:17.

CHAPTER THREE: THE INSTITUTES OF CHEMISTRY

1. Res. Cur., 4 January 1702, in Molhuysen, *Bronnen*, 4:190.

2. On the problem of "analysis by fire," see Debus, "Fire Analysis."

3. Other historians, notably Rosaleen Love, have called Boerhaave's scheme the "element-instrument" theory. I prefer "instrument" theory because (1) according to Boerhaave, the instruments do not function as "elements," i.e., first principles—they are tools; and (2) one of the five instruments, chemical menstrua, was never an "element" in any system. See Love, "Herman Boerhaave."

4. Boerhaave, *Oratio de Commendando Studio Hippocratico*. The English quotations are from Boerhaave, "Oration to Recommend the Study of Hippocrates," in *Boerhaave's Orations*, 54–84, at 79, 83.

5. On the goal of medical education at this time, see Frijhoff, *La Société neederlandaise*; Lindemann, *Medicine and Society*, 102–3; Brockliss and Jones, *Medical World*, 476.

6. See "Forma Examinis, qua Doctor Medicinae," in Molhuysen, *Bronnen*, 1:48*; Lindeboom, "Medical Education," 203.

7. This emphasis on theoretical knowledge was commonplace in university medical faculties. See, for example, the German case: Broman, *Transformation*, 29–30.

8. On the disputation exercise and its textual base, the dissertation, see Chang, "From Oral Disputation to Written Text."

9. Broman, *Transformation*, 32.

10. Boerhaave, *Institutiones Medicae*; Boerhaave, *Aphorismi*; Lindeboom, *Herman Boerhaave*, 70–78.

11. On the significance of method in medical courses, see Wightman, "*Quid sit Methodus?*"; Ong, *Ramus*.

12. See Boerhaave, *Institutiones Medicae*; Lindeboom, *Herman Boerhaave*, 70–78; Cook, "Boerhaave," 234–36.

13. Bohn, *Dissertationes*. The dates of the exercises and the name of each student *defendens* were printed at the beginning of each set of theses. On Bohn's career, see Rothschuch, "Bohn, Johannes."

14. "Ad Chymico Physicas Bohn Observationes," in VMA, MS 3, 94r–117r.

15. "Collegium Chemicum," in VMA, MS 3, 1r–93v.

16. For this historical introduction, see ibid., 2r–11v. An expanded version of this history can be found in Boerhaave, *Elements of Chemistry*, 1:4–17. An interpretation of this history is found in Christie, "Historiography of Chemistry."

17. See Hannaway, *Chemists and the Word*, 144–51.

18. "Collegium Chemicum," in VMA, MS 3, 15r: "Chemia est arts quae / docet / exercere operationes, quibus corpora, quae vasis capi aut coërceri possunt, solvuntur vel componuntur ope

quorundam instrumentorum ut corporum effectus & mutationes, medicamenta efficacia, tuta, grataque inveniantur reapse, utque variae artes commodius pertractentur."

19. Ibid., 23v–24r.

20. Ibid., 15r–22r.

21. See Lemery, *Cours de chymie*. Lemery first divided the operations according to their starting materials into "mineral," "vegetable," and "animal" substances. Under "minerals," the first class of operations involved gold, then silver, etc.

22. Ibid., 5–7; Stam, "proaemium," in Vigani, *Medulla Chymiae*, 7–8.

23. Boerhaave, *Elements of Chemistry*, 1:78–500. On Boerhaave's interest in fixed alkali, see 1:440–62. On "elective affinity," see Duncan, *Laws and Order.*

24. Debus, "Fire Analysis."

25. Boyle, *Sceptical Chymist*; Boyle, *Producibleness of Chymical Principles*; Boyle, *Experiments, Notes*. On Boyle's *Sceptical Chymist* and his critique of chemical principles, see Principe, *Aspiring Adept*, 27–62; Debus, "Fire Analysis."

26. "Collegium Chemicum," in VMA, MS 3, 24r: "Effecta separationis ad nullas similes partes, neque ad Elementa revocari queunt, neque sine partium immutatione semper producuntur sola separatione."

27. Boerhaave, *Elements of Chemistry*, 1:46. See also Love, "Herman Boerhaave," 549–51.

28. Boerhaave, *Elements of Chemistry*, 1:46.

29. On Bohn, see Rothschuch, "Bohn, Johannes"; Trevisani, "'Ratio' und 'Experimentum.'"

30. Bohn, "De Corporum Dissolutione," in *Dissertationes*, paras. 12–17. Each dissertation is unpaginated.

31. See Sennert, *De Chymicorum*, 386–87.

32. Newman, *Atoms and Alchemy*, 85–153.

33. On Rolfinck's career and work, see Partington, *History of Chemistry*, 2:312–14.

34. Rolfincius, *Chemia*, 37–74; Sennert is discussed at 42. On Van Helmont and the principles, see Debus, "Fire Analysis," 137–38.

35. Rolfincius, *Chemia*, 105–14.

36. See Partington, *History of Chemistry*, 4:653–86, at 653.

37. I have used the following English translation of Stahl's *Fundamenta*: Stahl, *Philosophical Principles*, 58–67. On Stahl's early lectures, and his textbook, see Olroyd, "An Examination."

38. "Collegium Chemicum," in VMA, MS 3, 28r: "Inter lucentes & urentes simul primo est ignis solis perennis, penetrantissimus, purissimus sine ulla faece. Constant in celerrina' attritione partium solidarum." See Bohn, "Dissertationum Chymico-Physicarum Tertia: De Igne," in *Dissertationes*, 3–4. See also Love, "Some Sources."

39. "Collegium Chemicum," in VMA, MS 3, 28v.

40. Bohn, "Dissertationum Chymico-Physicarum Tertia: De Igne," in *Dissertationes*, 6.

41. Senguerd, *Philosophia Naturalis*, 340–42. Rosaleen Love has argued that Boerhaave's concept of fire derived from Cartesian ideas about active "first matter." Although Boerhaave's fire could be placed within this intellectual pedigree, I focus on the direct source of his ideas; she mentions neither Bohn nor Senguerd in her work. See Love, "Herman Boerhaave"; Love, "Some Sources."

42. Senguerd, *Philosophia Naturalis*, 342.

43. "Collegium Chemicum," in VMA, MS 3, 28r–v; Love, "Some Sources," 165–72. On Duclos's and Homberg's work, see Duclos, "De l'augmentation du poids," 34–37; Duclos, "Dissertation sur les principles des mixtes naturels," 1–40; Homberg, "Observations faites par le moyen du verre ardent"; Homberg, "Essais de chimie. Article premier," 44–48. Homberg published three more memoirs through 1707 on experiments with the Duke of Orleans' burning lens, after which his junior colleague, Étienne-François Geoffroy, took over the brunt of the research. On the use of burning lenses in eighteenth-century chemistry, including that of the Duke of Orleans, see Smeaton, "Some Large Burning Lenses."

44. Principe, "Wilhelm Homberg," 539–46; Homberg, "Observations faites par le moyen du verre ardent"; Homberg, "Essais de chimie. Article premier."

45. Chang, "Fermentation, Phlogiston and Matter Theory"; Eklund, "Chemical Analysis and the Phlogiston Theory"; Stahl, *Zymotechnia Fundamentalis*.

46. "Collegium Chemicum," in VMA, MS 3, 28v: "Quocunque motu excitatus ignis perit, nisi adsint corpora in quae agat, haec pabulum audiunt ignis . . . : quo suppeditatio conservatur, interditur atque augetur pro lubitar." See Bohn, "Dissertationum Chymico-Physicarum Tertia: De Igne," in *Dissertationes*, 5–6.

47. "Collegium Chemicum," in VMA, MS 3, 30v–31r.

48. See Boerhaave, *Elements of Chemistry*, 1:168–81.

49. "Collegium Chemicum," in VMA, MS 3, 33r–v: "[Äer] Constat corporibus minutissimis, mobilissimis, solidissimis."

50. De Pater, "Experimental Physics," 320–22. On Boyle's experimental work on the properties of "air" conducted with his air pump, see Shapin and Schaffer, *Leviathan and the Air-Pump*; Hall, *Robert Boyle*, chap. 3. Boerhaave also modeled his discussion of "air" on Johannes Bohn's work; cf. Bohn, "Dissertationum Physico-Chemicarum Quatra: De Aere," in *Dissertationes*.

51. "Collegium Chemicum," in VMA, MS 3, 31v: "Ignis äere orbatus, vel cum eodem diu concusus, extinguitur. Sed in äere viget, ejus motu augetur, cöercendo, movendo, atterendo."

52. Ibid., 33v.

53. Ibid., 36r–v. The notion of the "fixity" of salts in earth is vague in Boerhaave's early lecture notes but is much better described in the *Elementa Chemiae*; see Boerhaave, *Elements of Chemistry*, 1:371–72.

54. "Collegium Chemicum," in VMA. MS 3, 36v: "Menstruum dicitur quod applicatum corporibus ea dividit in partes ex quibus constabat tota sive fluida, sive solida, atque dein cum iis fluitat vel concrescit." For Bohn's definition and discussion of chemical menstrua, see Bohn, "De Menstruis," in *Dissertationes*.

55. Boerhaave mentions cinnabar in this regard in "Collegium Chemicum," VMA, MS 3, 36v. For an extended discussion of cinnabar in this context, see Boerhaave, *Elements of Chemistry*, 1:399–400. Boerhaave also mentions the alchemical context of this operation. The addition of iron filings to cinnabar produced (in modern parlance) iron sulfide or "fools gold." Boerhaave explained both how this process could be used by the unscrupulous to fool those unskilled in chemistry, but also how this suggested to "alchemists" that mercury and sulfur (with iron) could ultimately be used to make gold.

56. "Collegium Chemicum," VMA. MS 3, 41r.

57. Ibid., 38v.

58. Lemery, *Cours de chymie*, 22–25; Homberg, "Essais de chimie. Article premier," 54–56.

59. "Collegium Chemicum," VMA, MS 3, 38r: "Ergo inductione patet menstrua non agere nisi moveantur, mota tamen quaelibet menstrua non agunt in quaecunque, sed determinata menstrua in determinata quaedam objecta. Illa determinatio haeret in menstruo, in solvendo, vel in utrisque."

60. On the "dry" way and "wet" way in early modern chemistry, see Eklund, *Incompleat Chymist*, 6–7, 12–14; Holmes, "Analysis by Fire."

61. "Collegium Chemicum," in VMA, MS 3, 36v.

62. Ibid., 38r.

63. Ibid., 40r.

64. Ibid., 48v. For an extended discussion, see Boerhaave, *New Method of Chemistry*, 2:7.

65. "Collegium Chemicum," in VMA, MS 3, 50r.

66. Ibid., 50r–v. Boerhaave cites Boyle's "de Util. Phil. Exp." (50v). Cf. Boyle, *Some Considerations*.

67. "Collegium Chemicum," in VMA, MS 3, 50v. On the increasing use among chemists at this time of solvent extraction, of which Boerhaave was an influential advocate, see Holmes, "Analysis by Fire."

68. "Collegium Chemicum," in VMA, MS 3, 55r–58r. For an in-depth examination of this sequence of operations, see Klein, "Experimental History." On "traditional" distillation analysis, see Multhauf, "Significance of Distillation." For the eighteenth century, see Holmes, *Eighteenth-Century Chemistry,* chap. 2.

69. "Collegium Chemicum," in VMA, MS 3, 57r–60r.

70. Ibid., 60r–v: "Usus optimus est, ut dissoluta aliquat grana, v.g. pro adulto XX, in aquae purae [oz.] X hauriantur iejuno stomacho divisa in tres partes aequales, quarum singula potetur calida, interposita inter binas doses 1â/2 hora & leni ambulatione."

71. Pagel, *Joan Baptista van Helmont,* 80 and passim; Newman, *Gehennical Fire,* 143–46, 148–51. For the development of Van Helmont's notion of fermentation in the later seventeenth century, see Anna Marie Roos's examination of William Simpson, in Roos, *Salt of the Earth,* 111–31.

72. See Bohn, "De Fermentationes," in *Dissertationes.* On Stahl, see Chang, "Fermentation, Phlogiston and Matter Theory."

73. VMA, MS 6, 91r–94v.

74. "Collegium Chemicum," in VMA, MS 3, 75r: "Eadem vegetibilia arte quadam mutari possunt in liquidum tenue, volatile, limpidum, acre, inflammabile [i.e., alchohol]; vel in liquidum tenue, minus volatile, limpidum, acidum, non flammable [i.e., vinegar]: operation haec fermentatio appellatur."

75. Ibid., 76v–80v. Quote at 76v: "6. Aquas ex observatione Britannorium fluviales in spiritum ardentem fatiscentes post sponte conceptam fermentationem."

76. VMA, MS 3, 143r.

77. Klein, "Experimental History."

78. Boerhaave, *Elements of Chemistry,* 2:1.

79. Ibid., 2:1–2.

80. VMA, MS 3, 143r (operations 1–6). Later in the *Elementa,* Boerhaave describes the "gentle" heat of the first operation as no more than 85 degrees Fahrenheit; see Boerhaave, *Elements of Chemistry,* 2:9.

81. Boerhaave states that this same set of operations can be used on a number of plant substances; see Boerhaave, *Elements of Chemistry,* 2:11–12.

82. Ibid., 2:2.

83. Ibid., 2:39–40. Boerhaave distinguished several different types of alkali salt based on the method of synthesis and their subtly differing chemical properties, i.e., some were more "acrid" than others.

84. Ibid., 2:44.

85. The operation lists for Boerhaave's demonstration courses are found in VMA, MS 3, 143r–78v.

86. Ibid., 26r–27v. See chap. 5 on the effects that Fahrenheit's thermometer had on Boerhaave's chemistry.

87. Boerhaave recorded the start date each time he taught his lecture course; see VMS, MS 3, 1r.

88. See Lindeboom, *Herman Boerhaave,* 109–10.

89. Wolfganck, *Disputatio Chymico-Medica*; Snellen, *Disputatio Metallurgico-Physico-Medica.*

90. See Le Mort, *Chymia Verae Nobilitas.*

CHAPTER FOUR: CHEMISTRY IN THE MEDICAL FACULTY

1. Boerhaave, *Sermo Academicus de Chemia suos Errores Expurgante.* English quotations are from Boerhaave, "Discourse on Chemistry Purging Itself of Its Own Errors," in *Boerhaave's Orations,* 180–213, at 207.

2. Boerhaave, "Chemistry Purging Itself," in *Boerhaave's Orations,* 208.

3. Boerhaave, *Dr. Boerhaave's Academical Lectures,* 1:268–69.

4. See, for example, Debus, *Chemistry and Medical Debate*, 184–207; Lindeboom, "Boerhaave's Concept."

5. See Boerhaave's *De Usu Ratiocinii Mechanici in Medicina*. An English translation may be found in Boerhaave, "Oration on the Usefulness of the Mechanical Method in Medicine," in *Boerhaave's Orations*, 85–120.

6. Knoeff, "Chemistry, Mechanics, and the Making of Anatomical Knowledge"; Knoeff, *Herman Boerhaave*.

7. See Guerrini, "Varieties of Mechanical Medicine." Luyendijk-Elshout has characterized Boerhaave's medicine as "mechanical" in order to distinguish his approach from that of the eighteenth-century "vitalists" (like Stahl). Within this context, "chemical" properties could be seen as "mechanical" (or not), depending on how one understood chemical action. See Luyendijk-Elshout, "Mechanisme contra vitalisme."

8. Boerhaave, "Oration to Recommend the Study of Hippocrates," in *Boerhaave's Orations*, 54–84, at 79, 83.

9. See Meinel, "*Artibus Academicis Inserenda*"; Lindeboom, "Boerhaave's Impact." In most medical faculties, chemistry was taught by private lecturers outside the university. In the case of Britain, see Guerrini, "Chemistry Teaching."

10. On Paracelsian chemical medicine and its conflicts with Galenic, academic medicine, see Pagel, *Paracelsus*; Pagel, *Joan Baptista van Helmont*; Debus, *French Paracelsians*; Trevor-Roper, "Paracelsian Movement."

11. Debus, *French Paracelsians*, chaps. 2–3; Brockliss and Jones, *Medical World*.

12. Clericuzio, "Teaching Chemistry."

13. Moran, *Chemical Pharmacy*.

14. An excellent discussion of how these traditions came together in Sylvius's experimental work is found in Ragland, "Experimenting with Chymical Bodies," 619–30.

15. Debus, *Chemistry and Medical Debate*, 59–64. On Van Helmont's theory of digestion, and acids and alkalis, see ibid., 43–47; Pagel, *Joan Baptista Van Helmont*, 129–40.

16. On the varieties of the acid/alkali theory, see Hall, "Acid and Alkali."

17. Sylvius, *Opera Medica*, 158. On Sylvius's medical system, see Beukers, "Acid Spirits and Alkaline Salts"; Beukers, "Mechanische Principes"; Baumann, *François Dele Boë Sylvius*; King, *Road to Medical Enlightenment*, 93–113; Debus, *Chemistry and Medical Debate*, 59–64; Partington, *History of Chemistry*, 2:281–90; Underwood, "Franciscus Sylvius."

18. Smith, "Science and Taste."

19. Ragland, "Experimenting with Chymical Bodies." The best overview of Sylvius's anatomical research remains Foster, *Lectures*, originally published in 1901.

20. Cook, *Trials of an Ordinary Doctor*, 63. On Sylvius's popularity, see Lindeboom, "Medical Education," 206; Cook, *Matters of Exchange*, 149–50. Also see Sylvius's request to expand the dissection theater at the hospital to accommodate the growing number of students: Res. Cur., 8 February 1660, in Molhuysen, *Bronnen*, 3:158.

21. Eklund, *Incompleat Chymist*, 14.

22. Lindeboom, "Frog and Dog," 283.

23. Ibid., 289–90. For Drélincourt's *praxis medica* course, see British Library, Sloane Collection, MS 1268; for student notes on physiological demonstrations, see Sloane Collection, MS 1291, 1r–38r.

24. Lindeboom, "Frog and Dog," 290–91; Ruestow, "Doctrine of Vascular Secretion," 267.

25. Cook, *Matters of Exchange*, 268–88.

26. On Willis, see Debus, *Chemistry and Medical Debate*, 64–73; Bynum, "Anatomical Method"; Hughes, *Thomas Willis*.

27. See British Library, Sloane Collection, MS 1268, 17r–266v. These remedies, associated with the Paracelsian tradition, had been the subject of intense debate earlier in the seventeenth century; see Debus, *French Paracelsians*, 95–99.

28. Nuck, *De Ductu*.

29. Ibid., praefatio.

30. Wear, "William Harvey." See also Cook, *Matters of Exchange*, chap. 10.

31. Res. Cur., 8 August 1718, in Molhuysen, *Bronnen*, 4:303: "Professor Boerhaaven sijnde bekent gemaekt dat het de dewoonte was in diese Universiteyt, dat de respective Professoren voor het aenvaaren van haaren dienst een publycque oratie doen, omme by die occasie de nuttigheyt van haare aanbevole institutien aan te mijsen ende de studenten tegelijck tot naarstigheit te encourageeren."

32. Boerhaave, "Chemistry Purging Itself," in *Boerhaave's Orations*, 206–7.

33. Boerhaave, "Chemistry Purging Itself," in *Boerhaave's Orations*, 197. The reference was to the Paracelsian work *De Nymphis, Sylphis, Pygmaeis et Salamandris et Caeteris Spiritibus*, which was published several times during the late sixteenth and early seventeenth centuries. See Paracelsus, *Sämtliche Werke*, 14:115–51. For an English translation, see Paracelsus, *Four Treatises*, 223–53.

34. Boerhaave, "Chemistry Purging Itself," in *Boerhaave's Orations*, 197. Later in the oration, Boerhaave held up Robert Boyle as the ideal chemist, stating that he distinguished "in the most prudent matter between the principles of religion and the whole field of natural and chemical science" (198).

35. Ibid., 193–94.

36. Ibid., 207–8.

37. Ibid., 201, 202.

38. Ibid., 202.

39. Ibid., 210, 202.

40. Rina Knoeff has explored in detail the connections between Boerhaave's "Hippocratic" medicine and his chemistry; see Knoeff, "Practicing Chemistry."

41. In this context, "method" was also called *ordo*. See Gilbert, *Renaissance Concepts of Method*; Dear, "Method and the Study of Nature"; Wightman, "*Quid sit Methodus?*"

42. Boerhaave, *De Usu Ratiocinii Mechanici in Medicina*. I have cited Boerhaave, "Oration on the Usefulness of the Mechanical Method in Medicine," in *Boerhaave's Orations*, 85–120.

43. "Praelectiones de Methodo Addischendae Medicinae," VMA, MS 26.

44. Bennet, "Mechanics' Philosophy"; Bennet, "Robert Hook."

45. This approach to natural philosophy, also called "physico-mathematics," had been gaining popularity throughout the seventeenth century; see Dear, *Discipline and Experience*, 210–27. Regarding the idea of "constructive geometry" in mathematical practitioners and natural philosophers, see Garrison, "Newton."

46. Boerhaave, "Mechanical Method in Medicine," in *Boerhaave's Orations*, esp. 117–19; Boerhaave, "Discourse on the Achievement of Certainty in Physics," in *Boerhaave's Orations*, 145–79, at 175–77. This second oration was delivered in 1715. For an outline of the "method of mechanics" in Boerhaave's institutiones medicae course, see Boerhaave, *Dr. Boerhaave's Academical Lectures*, 1:57–76.

47. Ruestow, *Physics*, 105–10.

48. See Domenico Bertoloni Meli's work on collaborations between physicians and mathematicians: Bertoloni Meli, "Collaboration."

49. See Guerrini, "Tory Newtonians"; Guerrini, "Archibald Pitcarne"; Brown, "Medicine in the Shadow of the *Principia*."

50. Guerrini, "Archibald Pitcarne." The possible connection between Boerhaave and Pitcarne has been a contentious issue; cf. Lindeboom, "Pitcarne's Leyden Interlude."

51. "Praelectiones de Methodo Addischendae Medicinae," VMA, MS 26, 2v–5r, 7r.

52. Ibid., 7r–8r. Mathematical training was so important for Boerhaave's medical system that he offered private medical courses in "mechanics" for medical students; see his course from 1710–11: "De His, Quae Medico ex Mechanicis, Hydrostaticis atque Hydraulicis Scitu Necessaria," VMA, MS 25.

53. On eighteenth-century notions of "physics," see Heilbron, *Elements of Early Modern Physics*, chap. 1; Schaffer, "Natural Philosophy" (1980).

54. De Pater, "Experimental Physics"; Wiesenfeldt, *Leerer Raum,* chap. 3. See also Ruestow, *Physics,* 96–104. See also notes from De Volder's experimental course, in the British Library, Sloane Collection, MS 1292.

55. "Praelectiones de Methodo Addischendae Medicinae," VMA, MS 26, 9r, 9v.

56. Ibid., 11r.

57. Ibid.

58. See Debus, "Fire Analysis"; Holmes, "Analysis by Fire." On the importance of distillation in eighteenth-century chemistry, especially in the analysis of plant and animal matter, see Holmes, *Eighteenth-Century Chemistry,* chap. 3.

59. For an excellent discussion of how eighteenth-century chemists used reagents with known properties, often through complex operations, to test unknown substances, see Eklund, *Incompleat Chymist,* 15–19. On the origins of this type of analysis from the metallic assaying tradition, see Greenaway, "Early Development of Analytical Chemistry," 91–97. See also the "Helmontian" roots of this practice: Newman and Principe, *Alchemy Tried in the Fire,* chaps. 2–3.

60. "Praelectiones de Methodo Addischendae Medicinae," VMA, MS 26, 11r.

61. Boerhaave, *Institutiones Medicae.* Boerhaave published revised editions in 1713, 1720, 1727, and 1735. See Lindeboom, *Bibliographia Boerhaaviana.* I have used an English translation of the *Institutiones Medicae,* probably from the 1727 edition; Boerhaave, *Dr. Boerhaave's Academical Lectures.*

62. Boerhaave, *Dr. Boerhaave's Academical Lectures,* 1:268.

63. Ibid., 1:280.

64. One prime example was the work of Robert Boyle; see Knight and Hunter, "Robert Boyle's *Memoirs*"; Büttner, "Die physikalische und chemische Untersuchung." Even Carel de Maets, Leiden's first professor of chemistry, included in his course an operation to obtain the "salt of blood," which was (following Sylvius) an acid; see VMA,MS 131, 136r.

65. Stolberg, "Decline of Uroscopy"; Wall, "Inventing Diagnosis"; Shackelford, "Paracelsian Uroscopy."

66. Newman and Principe, *Alchemy Tried in the Fire,* 277–81.

67. See, for example, Nicolas Lemery's long discussion of urine and phosphorus: Lemery, *Cours de chymie,* 799–851.

68. Boerhaave, *Dr. Boerhaave's Academical Lectures,* 3:136. See also, on the production of the "natural salt" of urine, Boerhaave, *Elements of Chemistry,* 2:317–18.

69. Boerhaave, *Dr. Boerhaave's Academical Lectures,* 3:138.

70. Ibid., 3:137. Aqua regia was an acidic menstruum, made from a mixture of spirit of niter (nitric acid) and spirit of sea salt (hydrochloric acid), that could dissolve gold, whereas its two component acids by themselves could not. That he could make an aqua regia from urine even after he had putrefied (thus, altering) the urine suggested to Boerhaave that spirit of salt (produced from table salt) must be present. Van Helmont had earlier discovered the presence of table salt in human urine, but Boerhaave did not mention any awareness of this fact; cf. Newman and Principe, *Alchemy Tried in the Fire,* 286.

71. The experiment in question occurred from July to October 1696 and was part of a series of experiments in which Boerhaave was testing for acids that could dissolve gold. Hollandus, whom Boerhaave frequently cited in his chemical work, had written treatises on the transformative power of urine. Unfortunately, Boerhaave's notes on the experiment do not mention Hollandus or his inspiration for the experiment. See VMA, MS 1, 18v. On Hollandus and urine, see Hollandus, *Tractatus de Urine.*

72. Luyendijk-Elshout, "*Oeconomia Anamalis.*" On the "animal oeconomy" more generally in seventeenth-century medicine, see Brown, *Mechanical Philosophy and the "Animal Economy."* Boerhaave titled the first and largest section of the *Institutiones Medicae,* in which he described his system of physiology, "On the Animal Oeconomy"; see Boerhaave, *Dr. Boerhaave's Academical Lectures,* vols. 1–5.

73. On the "vascular" model of the body, see Ruestow, "Doctrine of Vascular Secretion"; Knoeff, "Chemistry, Mechanics, and the Making of Anatomical Knowledge."

74. Boerhaave, *Dr. Boerhaave's Academical Lectures*, 5:277–302.

75. King, *Medical World of the Eighteenth Century*, 74–92. King's work remains the best, most concise survey of Boerhaave's theory of disease. See also Jevons, "Boerhaave's Biochemistry," 353–56. For Boerhaave's general discussion of diseases caused by corrupted humors, see Boerhaave, *Dr. Boerhaave's Academical Lectures*, 6:167–78; on "inflammation" and "fever," respectively, see Boerhaave, *Boerhaave's Aphorisms*, 83–92, 127–33.

76. Cf. Galen, *De Febrium Differentiis*, in *Medicorum Graecorum Opera*, 273–405, esp. 294–300; Galen, *De Caussis Morborum Liber*, ibid., 1–41. On Galen's physiology and pathology, see Nutton, *Ancient Medicine*, 230–47; Johnston, *Galen*, 3–126.

77. Pagel, *Paracelsus*, 153–61.

78. Boerhaave, *Dr. Boerhaave's Academical Lectures*, 1:184–89; Boerhaave, *Elements of Chemistry*, 2:93–99.

79. King, *Medical World of the Eighteenth Century*, 78–83; Jevons, "Boerhaave's Biochemistry," 353–56; Boerhaave, *Boerhaave's Aphorisms*, 16–18; Boerhaave, *Elements of Chemistry*, 1:425.

80. Boerhaave, *Elements of Chemistry*, 1:424.

81. See Boerhaave, *Boerhaave's Aphorisms*, 387–91. See also Boerhaave's private lecture course on stones, which he gave in June and July 1729: "Praelectiones de Calculo," VMA, MS 30, 1r–10r.

82. King, *Medical World of the Eighteenth Century*, 83–84.

83. Boerhaave, *Dr. Boerhaave's Academical Lectures*, 1:269–76.

84. King, *Medical World of the Eighteenth Century*, 80; Boerhaave, *Boerhaave's Aphorisms*, 17–18. "Crab's eyes" are calcareous masses that formed in the abdomens of crayfish and were ground and used as absorbent alkalis in pharmacy.

85. Jevons, "Boerhaave's Biochemistry," 354; Boerhaave, *Elements of Chemistry*, 1:42–43, 2: 93–99.

86. VMA, MS 28, 7r–v, 9r–v.

87. Ibid., 9r.

88. Boerhaave, *Elements of Chemistry*, 1:44–47.

89. Ibid., 1:52.

90. On the dissemination of the "Leiden model," see Lindeboom, *Herman Boerhaave*, 360–74. Versions of De Methodo transcribed from student notes were often published in editions of Boerhaave's *Opera* and were often included as addenda to the *Institutiones Medicae*, thus presenting an authoritative account of the "Leiden model"; see Lindeboom, *Bibliographia Boerhaaviana*.

CHAPTER FIVE: INSTRUMENTS AND THE EXPERIMENTAL METHOD

1. Res. Cur., 24 June 1718, in Molhuysen, *Bronnen*, 4:302.

2. Lindeboom, *Herman Boerhaave*, 109–10.

3. On the Leiden physicists, see Wiesenfeldt, *Leerer Raum*; De Pater, "Experimental Physics." On collaborations between physicists and instrument makers, see Van Helden, "Theory and Practice."

4. On Boerhaave's use of "Baconian" experimentalism, see Klein, "Experimental History." For the Dutch context, see also Elena, "Baconianism in the Seventeenth-Century Netherlands." On "Baconian" aims, experimentalism, and methodologies generally, see Webster, *Great Instauration*; Sargent, "Scientific Experiment and Legal Expertise"; Gaukroger, *Francis Bacon*; Pérez-Ramos, *Francis Bacon's Idea of Science*.

5. Dear, "Method and the Study of Nature"; Jardine, "Epistemology of the Sciences"; Wallace, *Causality and Scientific Explanation*, vol. 1; Gilbert, *Renaissance Concepts of Method*; Ong, *Ramus*. I

use the term "discipline" here in the early modern sense of *disciplina,* meaning "a learned subject"; see Ong, *Ramus,* 156–57, 163.

6. Bensaude-Vincent and Blondel, *Science and Spectacle*; Roberts, "Going Dutch"; Stewart, *Rise of Public Science*; Golinski, *Science as Public Culture*; Schaffer, "Consuming Flame"; Schaffer, "Natural Philosophy and Public Spectacle." On public anatomy demonstrations, see Guerrini, "Anatomists and Entrepreneurs"; Rupp, "Matters of Life and Death."

7. Shapin, "House of Experiment."

8. On Fahrenheit's lecturers, see Cohen and Cohen-de Meester, "Daniel Gabriel Fahrenheit," 14–17. On Musschenbroek, see De Clercq, *Sign of the Oriental Lamp*; Van Helden, "Theory and Practice."

9. VMA, MS 7. Boerhaave did not lecture during the 1722–23 winter term because he was ill; see Lindeboom, *Herman Boerhaave,* 126–28.

10. Cf. Lindeboom, *Herman Boerhaave,* 281–22; Burton, *Account of Boerhaave,* 167. For some of Boerhaave's private courses, see, for example, "Praelectiones de Calculo," in VMA, MS 30; Schulte, *Hermanni Boerhaave Praelectiones de Morbis Nervorum.*

11. VMA, MS 7, 1v.

12. Boerhaave, *Sermo Academicus de Comparando Certo in Physicis.* I quote the English text "Oration on the Achievement of Certainty in Physics," in Boerhaave, *Boerhaave's Orations,* 143–79, at 177–78.

13. See, for example, Bacon, "Great Instauration," in *New Organon,* 1–29, esp. 19–24; Klein, "Experimental History." On Baconian fact collecting and induction, see Farrington, *Philosophy of Francis Bacon*; Webster, *Great Instauration.*

14. Boerhaave, "Oration to Recommend the Study of Hippocrates," in *Boerhaave's Orations,* 54–84.

15. Rina Knoeff has pointed this out as well, in Knoeff, "Practicing Chemistry," 68.

16. Boerhaave, "Oration on the Achievement of Certainty in Physics," in *Boerhaave's Orations,* 162–63. On the much more complicated intricacies of Newton's scientific method, a good place to start is Guerlac, "Newton and the Method of Analysis."

17. "Praelectiones de Methodo Addischendae Medicinae," VMA, MS 26. Boerhaave's use of the term "physics" here refers to the common usage at that time—the study of natural bodies and motion—not the experimental physics course in Boerhaave's curriculum.

18. VMA, MS 7, 1v: "Communes omnibus habentur extensio, soliditas, gravitas, figurabilitas, divisibilitas, mobilitas, unitas quaedam elementorum cohaerentium."

19. Ibid., 2r: "Errant interim Mathematici, si putant hac via totam naturam exhauriri posse, quasi praeter generalia haec, non essent alia in corporibus, & quidem aeque efficacia."

20. Ibid.

21. Ibid.: "Cujus [chemiae] itaque utilitas ingens, absoluta necessitas, intelligitur."

22. Ruestow, *Physics,* 16–17; Reif, " Textbook Tradition," 28–30.

23. Roberts, "Going Dutch," 364–66; Wiesenfeldt, *Leerer Raum,* 108–11; Ruestow, *Physics,* 96–112.

24. Dear, "Narratives, Anecdotes, and Experiments"; Dear, *Discipline and Experience.*

25. Lawrence, "Educating the Senses." See also, regarding chemistry demonstrations after Boerhaave, Lehman, "Between Commerce and Philanthropy"; Roberts, "Chemistry on Stage."

26. For De Volder, see British Library, Sloane Collection, MS 1292, 78r–141v. For Newton, see Newton, *Opticks.*

27. VMA, MS 7, 4r.

28. Ibid., 3r–v.

29. Ibid., 4r.

30. Ibid., 4v. Boerhaave does not describe the thermometer in the lecture notes, but when he addressed this demonstration in the *Elementa Chemiae,* he described a Drebbel-type instrument. See Boerhaave, *Elements of Chemistry,* 94–95. Boerhaave purchased this thermometer from Daniel

Fahrenheit, and the thermometer tube was graduated according to Fahrenheit's scale. On this type of thermometer, see Middleton, *History of the Thermometer*, 19–21; Taylor, "Origin of the Thermometer," 151–56.

31. VMA, MS 7, 4v–5r.

32. Ibid., 5r. Boerhaave's number for the expansion of mercury was incorrect here. The correct figure should have been 1/54 (212 parts expanded in the tube/11,520 parts in the bulb).

33. Ibid., 5r: "ignis . . . vim habet ita rarefaciendi onmia corpora. De quo scitur illa proprietas, quod semper, ubique, sit: id autem patet sequentibus experientis."

34. Ibid., 5r, 5v.

35. Boerhaave, "Oration on the Achievement of Certainty in Physics," in *Boerhaave's Orations*, 158–59. On Bacon's arguments, see Gaukroger, *Francis Bacon*, 181–88.

36. VMA, MS 7, 4v: "Patet & inde cur pendula intra tropicos Longiora fiant, hacque cause tardius vibrentur?" The point Boerhaave was making in this example was that the hotter temperatures in the tropics would increase the length of a pendulum, thus increasing its period of oscillation. So it would take a longer time to swing out and back to its starting position.

37. Ibid.,55r–57v. The experiment was described at 57r.

38. Ibid., 61v–74r.

39. Cf. Duncan, *Laws and Order*, 57–59.

40. Cf. "1719 Sept. Series Lectionum" and "1720 Sept. Series Lectionum," in Molhuysen, *Bronnen*, 4:153*, 154*.

41. "Instructie ofte reglement by de H. C. ende B. beraemt ende nedergestelt voor de knegt van het Laboratorium Chymicum ende waarnaar hy sigh punctuelyck sal hebben gedragen," 26 September 1718, in Molhuysen, *Bronnen*, 4:150*.

42. See Res. Cur., 18 June 1674, in Molhuysen, *Bronnen*, 4:290–1. This resolution was originally issued against "Cartesian" students, who disrupted classes being taught by "Aristotelian" faculty members.

43. Shapin, "House of Experiment," 399–404. See also, more generally, Collins, "Public Experiments."

44. Boerhaave's remarks were recorded in the *Elementa Chemiae*; see Boerhaave, *Elements of Chemistry*, 1:222. The original demonstrations were found in VMA, MS 7, 15r–18r.

45. VMA, MS 7, 6v, 7v–8r. On these instruments, see Smeaton, "Some Large Burning Lenses."

46. VMA, MS 7, 57r–v.

47. VMA, MS 3, 143r–60v.

48. Boerhaave's use of instruments reflects Peter Galison's contention that instruments "embody continuities of practice, expertise, and objects" that are "partially autonomous" from theories and experimental programs. See Galison, "History, Philosophy, and the Central Metaphor."

49. "Collegium Chemicum," VMA, MS 3, 28r; Love, "Some Sources." On Senguerd's notion of "fire," see "Instruments of Chemistry," in chap. 3.

50. Middleton, *History of the Thermometer*, 41–79; Chang, *Inventing Temperature*, chaps. 2–3. Francis Hauksbee the younger's thermometers disagreed by as much as 10°F at lower temperatures; see Middleton, *History of the Thermometer*, 59. By contrast, Christiaan Wolff reported that two Fahrenheit spirit thermometers agreed to within 1/16° (on an earlier scale, equivalent to 1/4°F); see Middleton, *History of the Thermometer*, 74, and [Wolff], "Relatio de Novo Barometrorum."

51. Van der Star, *Fahrenheit's Letters*, 72–99.

52. Fahrenheit to Boerhaave, 12 December 1718, in Van der Star, *Fahrenheit's Letters*, 85. Boerhaave calculated the ratio for the spirit thermometer himself, and his figure differs significantly from Fahrenheit's figure of 96/1,675. Fahrenheit himself, however, was still in the process of establishing a stable value for the expansion of both spirit and mercury. For example, in 1718 Fahrenheit reported the expansion of mercury over 96° to be 1/120, but by the next year (1719) he reported a figure of 1/(115 7/8), and by 1729 he reported a figure of 1/(116 9/16).; see Van der Star,

Fahrenheit's Letters, 95, 139. On Fahrenheit's thermometers and method of graduation, see ibid., 18–31; Middleton, *History of the Thermometer,* 66–79.

53. Fahrenheit to Boerhaave, 12 December 1718, 85–87; Fahrenheit to Boerhaave, 23 January 1719, in Van der Star, *Fahrenheit's Letters,* 97–99.

54. Fahrenheit to Boerhaave, 12 December 1718, , 89.

55. VMA, MS 7, 15v–18r.

56. Fahrenheit to Boerhaave, 12 December 1718, 88–91. Boerhaave's program follows one that Francis Bacon suggested in book 2 of the *New Organon;* see Bacon, *New Organon,* 131–32, 140–41.

57. VMA, MS 7, 15v.

58. Boerhaave, *Elements of Chemistry,* 1:222.

59. De Pater, "Experimental Physics"; Wiesenfeldt, *Leerer Raum;* Ruestow, *Physics.*

60. VMA, MS 7, 22r: "Spatia occupata ab eadem portione aeris esse in ratione reciproca ponderum comprimentum."

61. For example, 4,224 lbs. of weight reduced the volume to half the original volume; 8,226 lbs. reduced the air to one-fourth its volume; etc. See VMA, MS 7, 21v.

62. Ibid., 24v–27v.

63. Van Helmont, "Gas Aquae," in *Ortus Medicinae,* 71–81; Pagel, *Joan Baptista van Helmont,* 60–70. Paraphrased from Van Helmont, "Progymnasma Meterori," in *Ortus Medicinae,* 73: "Gas, vapore, fuligine, & stillatis oleositatibus, longe sit subtilius, quamquam multoties aere adhuc densius."

64. Hall, *Robert Boyle,* 183–84, 187.

65. VMA, MS 7, 28v.

66. Hall, *Robert Boyle,* 187–96.

67. VMA, MS 7, 31r. Boerhaave did not record the modifications to his air pump in his lecture notes. They are found in Boerhaave, *Elements of Chemistry,* 1:308–9.

68. VMA, MS 7, 31r–v.

69. See "Instruments of Chemistry," in chap. 3; "Collegium Chemicum," VMA, MS 3, 33r–v.

70. VMA, MS 7, 31r: "Inde & apparet effervescentias oriri ab ipsa vi corporum effervescentium & eo ipso aërem eliminantium." This statement was sufficiently vague so that the air being "eliminated" could be either concreted or dissolved!

71. Ibid.,31r–v. Quotation at 31v: "8. Ac omnis effervsecentim fit, dum corpora amica ruunt in amplexus mutuos, expellint aerem intime unitum elementis corporum, faciantque [illegible word] aerem." I have deduced that this corollary was added after the rest of the lectures on air were composed because it was squeezed in between the previous corollary and Experimentum 10 in the manner of a margin note.

72. Donovan, *Philosophical Chemistry,* 131–33.

73. Van Swieten graduated from the Leiden medical faculty in 1725 yet remained in Leiden and attended Boerhaave's lectures until the latter's death in 1738. Van Swieten later went on to become an Imperial court physician in Vienna and to write an extensive commentary on Boerhaave's *Aphorisms.* See Lindeboom, *Herman Boerhaave,* 191–92, 226–27.

CHAPTER SIX: PHILOSOPHICAL CHEMISTRY

1. Boerhaave, *Elements of Chemistry,* 1:50–51, long quotation at 51; ibid., 2:2.

2. Ibid., 1:398.

3. Shaw, *Chemical Lectures,* 1. For the context of "philosophical chemistry," see Donovan, *Philosophical Chemistry;* Taylor, "Unification Achieved"; Golinski, "Peter Shaw"; Gibbs, "Peter Shaw." See also this approach to chemistry in Holmes, *Eighteenth-Century Chemistry.*

4. Boerhaave, *Elements of Chemistry,* 1:vii–x. On this episode, see also Lindeboom, *Herman Boerhaave,* 176–81; Gibbs, "Boerhaave's Chemical Writings."

5. On the culture and strategies of early modern book piracy, see Johns, *Nature of the Book,* 160–86, 444–75.

6. [Boerhaave], *Institutiones et Experimenta Chemiae*; Burton, *Account of Boerhaave,* 148.

7. Lindeboom, *Bibliographia Boerhaaviana,* 80–81, nos. 445–49; [Boerhaave], *New Method of Chemistry* (1727).

8. [Boerhaave], *New Method of Chemistry* (1727), 187.

9. Res. Cur., 1, 8, and 26 February 1729, in Molhuysen, *Bronnen,* 5:60, 69–70.

10. Boerhaave, "Academic Discourse, Delivered by Herman Boerhaave When He Officially Resigned His Professorships in Botany and Chemistry," in *Boerhaave's Orations,* 214–36, quotation at 233–34. On Boerhaave's illnesses, see Lindeboom, *Herman Boerhaave,* 126–28, 159–61.

11. Boerhaave, "Academic Discourse," in *Boerhaave's Orations,* 234.

12. On the editions of Aretaeus and Swammerdam, see Lindeboom, *Herman Boerhaave,* 184–86, 198–201. Boerhaave's chemical experiments with mercury are discussed in chap. 7.

13. The two chairs that Boerhaave relinquished were officially in botany and medicine, and chemistry and medicine. Thus, in effect, Boerhaave assumed a new chair, the one usually reserved for the professor of the institutes of medicine, although he never used this title. Because he was relieved of some of his duties, Boerhaave requested that his salary be reduced accordingly. The curators reduced Boerhaave's salary by 400 guilders at his request, but a year later they restored it to its previous level, 2,100 guilders a year. See Res. Cur., 8 August 1730, in Molhuysen, *Bronnen,* 5:94. On Boerhaave's course on nervous diseases, see Schulte, *Hermanni Boerhaave Praelectiones de Morbis Nervorum.*

14. On the final date of the instruments course, see VMA, MS 7, 75r; on the lecture course, see "Collegium Chemicum," in VMA, MS 3, 1r; on the operations course, see MS 3, 178 r-v.

15. Gibbs, "Boerhaave's Chemical Writings," 121–22; Boerhaave to Hans Sloane, 15 October 1731, in Lindeboom, *Boerhaave's Correspondence,* 1:196–97.

16. Lindeboom, *Herman Boerhaave,* 182, 344. Boerhaave used a similar strategy with his *Libellus de Materie Medica* (1719); he had the printer, Severnius, sign every copy. See Lindeboom, *Herman Boerhaave,* 118.

17. Lindeboom, *Herman Boerhaave,* 182–83. There is some debate about whether Imhoff presented the elector with a manuscript of the *Elementa,* as he also claimed in the "privilege."

18. Gibbs, "Boerhaave's Chemical Writings," 122–23.

19. Lindeboom, *Bibliographia Boerhaaviana,* 81–86, nos. 450–500; Gibbs, "Boerhaave's Chemical Writings," 131–35.

20. Boerhaave, *Elements of Chemistry,* 1:19.

21. Ibid., 1:1–3.

22. Earlier writers in natural philosophy shaped by scholastic pedagogy often claimed that arguments from experience were based on numerous observations or the common course of nature. See Dear, "Narratives, Anecdotes, and Experiments," 138; Dear, "Jesuit Mathematical Science." By the early eighteenth century (depending, of course, on the context and local philosophical culture), this argument had spread to experimental accounts. For example, Newton employed such "crucial" experiments in his *Opticks,* and in chemistry, Wilhelm Homberg (and the other chemists at the Académie des sciences) regularly represented their conclusions from diverse experiments in terms of one exemplary "operation." On Newton, see Dear, *Discipline and Experience,* 224–27, 233–43. On Homberg et al., see Holmes, "Argument and Narrative in Scientific Writing."

23. Debus, "Fire Analysis"; Holmes, "Analysis by Fire." For the *theoria* course, see chap. 3.

24. Boerhaave, *Elements of Chemistry,* 1:44–45.

25. Ibid., 1:46–47.

26. Ibid., 1:46.

27. "Air" was an exception to this rule; see "Hales on the Reactivity of Air."

28. Boerhaave, *Elements of Chemistry,* 1:19.

29. ibid., 1:168–201. On Boerhaave's conception of combustion see Love, "Herman Boerhaave."

30. Boerhaave, *Elements of Chemistry*, 1:170–81.

31. Ibid., 1:47, 181–82, 184–86. The products of alcohol combustion are water and carbon dioxide. Boerhaave's bell jar condensed the water, as it was designed to do, but he was unaware of the second, gaseous product.

32. Boerhaave, *Elements of Chemistry*, 1:186–91, quotation at 191.

33. For the *spiritus rector* experiments, see ibid., 1:47–49; on Boerhaave's "mercury" (the alchemical "mercury" principle), see chap. 7.

34. Boerhaave, *Elements of Chemistry*, 1:222, 292.

35. For example, in his section on the "objects" of chemistry, Boerhaave attached to the thesis that described the metals (no. 12), nine pages of additional notes on the properties of metals. These attached notes form the basis for the description of the metals in the *Elementa Chemiae*. "Collegium Chemicum," in VMA, MS 3, 15r–v (original thesis), 14v, 16r–22r (added notes); Boerhaave, *Elements of Chemistry*, 1:19–26.

36. See the additional notes contained in "Collegium Chemicum," in VMA, MS 3, 26r–27v.

37. Shapin, "Pump and Circumstance."

38. Boerhaave, *Elements of Chemistry*, 1:86.

39. Compare, for example, the first corollary in the instruments course—"Idem fere in omni solido corpore obtinet" (VMA, MS 7, 4r)—with the first corollary in the *Elementa*: "Now this expansion of solid Bodies by Fire is so universal, that I have never observed it to fail in any one that I have as yet an opportunity of making trail of" (Boerhaave, *Elements of Chemistry*, 1:86–87).

40. Boerhaave, *Elements of Chemistry*, 1:87. This incident was recorded in the correspondence between Boerhaave and Fahrenheit; Fahrenheit to Boerhaave, 30 March 1729, in Van der Star, *Fahrenheit's Letters*, 144–59.

41. Fahrenheit to Boerhaave, 20 March 1729, in Van der Star, *Fahrenheit's Letters*, 121–27; Boerhaave, *Elements of Chemistry*, 1:97–101. For the original *experimentum* in the instruments course, see VMA, MS 7, 4v.

42. Golinski, "Noble Spectacle."

43. Boerhaave, *Elements of Chemistry*, 1:222–25, quotation at 225. Boerhaave's complicated explanation for the spontaneous flaming of phosphorous was as follows: Recipes for phosphorous included animal parts ("juices," feces, or flesh), coal, and alum. Bodies that attracted water from the air, such as the oil of vitriol found in alum, often generated a great deal of heat through the attrition of air particles being dragged into the acid's pores along with the water. This heat could ignite a body that contained a *pabulum ignis* to sustain the flame, as the phosphorous did, since combustible coal was an ingredient in its preparation.

44. Several historians have recently made this claim; see Principe, *New Narratives*; Holmes, *Eighteenth-Century Chemistry*.

45. Kerker, "Herman Boerhaave"; Lavoisier, *Opuscules*, 26–28. On Hales's work on air, see Guerlac, "Continental Reputation of Stephen Hales." Hélène Metzger argued in 1930 that Boerhaave never accepted that air could be fixed. Henry Guerlac, citing Metzger, claimed that "this participation of air [in chemical reaction] was quite generally opposed in HALES' day on the Continent on the authority of such influential figures as BOERHAAVE and STAHL" (Guerlac, "Continental Reputation of Stephen Hales," 393). Guerlac, however, later argued that Boerhaave changed his opinion upon reading Hales; see Metzger, *Newton, Stahl, Boerhaave*, 246–59; Guerlac, *Lavoisier—The Crucial Year*, 25–27.

46. "Experimentum 9," in VMA, MS 7,31r–v; Boerhaave, *Elements of Chemistry*, 310–11.

47. Boerhaave, *Elements of Chemistry*, 1:312.

48. Ibid., 1:313, 314, 315.

49. For example, when Boerhaave poured spirit of niter upon iron filings in vacuo, he reported a large quantity of red fumes. He interpreted this phenomenon as the release of "elastic air" plus

the matter of the red fumes, not as one new type of air. In this instance, the air is performing its role as an instrument, acting as a medium for vapors and other airborne particles. See ibid., 1:312.

50. Ibid., 1:314.

51. Eklund, "Spirit in the Water"; Hales, *Vegetable Staticks*.

52. Fichman, "French Stahlism"; Rappaport, "G.-F. Rouelle," 94; Guerlac, "Continental Reputation of Stephen Hales."

53. See the classic study by Burnet, *Les physiciens hollandais*. Musschenbroek's interest in air (and in Hales) stemmed directly from Boerhaave's *Elementa*, and I suspect that Rouelle was led to Hales through Boerhaave as well. See Lehman, "Mid-Eighteenth-Century Chemistry."

54. Lavoisier, *Opuscules*, 11–25.

55. Lindeboom, *Herman Boerhaave*, 347.

56. Chang, "Fermentation, Phlogiston and Matter Theory." On "phlogiston" as "fixed fire," see Taylor, "Unification Achieved"; Kim, "'Instrumental' Reality of Phlogiston"; Siegfried, *From Elements to Atoms*, 100–113.

57. This comparison was based on Snelders, "Georg Ernst Stahls Phlogiston." Similar views are expressed in Metzger, *Newton, Stahl, Boerhaave*; Rappaport, "Rouelle and Stahl," 83–92; Melhado, "Oxygen, Phlogiston, and Caloric," 215–20.

58. King, "Stahl and Hoffman"; Geyer-Kordesch, "Passions and the Ghost in the Machine." On Stahl's Pietist faith, see Geyer-Kordesch, "Georg Ernst Stahl's Radical Pietist Medicine."

59. In the *Elementa*, Boerhaave recommended Hoffman's *Observationes Physico-Chemiae* (1722), and said: "[Hoffman] has done a vast deal of service to the chemical art, and enriched both Chemistry and Physic with [an] abundance of beautiful observations." See Boerhaave, *Elements of Chemistry*, 1:18. On Hoffman's philosophy, see King, "Medicine in 1695"; King, "Stahl and Hoffman"; and Lonie, "Hippocrates as Iatromechanist." On Boerhaave's reluctance to engage in public debates, see Lindeboom, *Herman Boerhaave*, 259–60.

60. Boerhaave, *Elements of Chemistry*, 1:326. Boerhaave rejected both of these claims.

61. Stahl, *Philosophical Principles*, 5–15; Oldroyd, "Examination." For Becher's system of elements, see Partington, *History of Chemistry*, 2:644–47. Stahl's earlier claim regarding water is found in Stahl, *Philosophical Principles*, 61.

62. "Natuurkundige Lessen van Daniel Gabriel Fahrenheit," Leiden University Archives, MS BPL 772, 84r; Geoffroy, "Eclaircissements sur la table," 211–19. Fahrenheit associated phlogiston with the "inflammable matter" (*verbrandelijk stoffen*) of oils, which "maintains and feeds the fire" (*vuur onderhouden en gevoed*). Geoffroy associated Stahl's phlogiston with his own (and Homberg's) "oily or sulpherous principle" (211, 213).

63. Recent historical work on phlogiston supports this view. Robert Siegfried, for example, suggests that "phlogiston" did not become a coherent theory until later in the century, when it was attacked by Lavoisier and his allies; see Siegfried, *From Elements to Atoms*, 110–13.

64. Boyle mixed oil of vitriol with oil of turpentine to synthesize sulfur, whereas Glauber fused charcoal and his *sal mirabile* (sodium sulfate) and added spirit of niter to the residue; see Eklund, "Chemical Analysis and the Phlogiston Theory," 11–12.

65. Homberg, "Essai de l'analyse du Souphre Commun"; Geoffroy, "Manière de recomposer le souffre commun." For the context of Homberg's experiments, see Principe, "Wilhelm Homberg."

66. Chang, "Fermentation, Phlogiston and Matter Theory." On Stahl's later work on sulfur, see Eklund, "Chemical Analysis and the Phlogiston Theory," 17–34.

67. Strube, "On the Importance"; Partington, *History of Chemistry*, 2:666–73.

68. "Collegium Chemicum," in VMA, MS 3, 28v–31r; Bohn, "Dissertationem Chymico-Physicarum Tertia: De Igne," in *Dissertationes*, 5–6 (each dissertation is individually paginated).

69. VMA, MS 7, 10v–11v; Boerhaave, *Elements of Chemistry*, 1:170–81.

70. Boerhaave, *Elements of Chemistry*, 1:186–91; VMA, MS 7, 12r.

71. VMA, MS 7, 12v; Boerhaave, *Elements of Chemistry*, 1:188–91, quotation at 191.

72. Stahl, *Zyomtechnia Fundamantalis,* in *Opusculum Chymico-Physico-Medicum,* 139 and passim. See also Chang, "Fermentation, Phlogiston and Matter Theory."

73. Boerhaave, *Elements of Chemistry,* 1:209–10, 201.

74. Partington, *History of Chemistry,* 2:670. This variance probably reflected the development of rather than an inconsistency in Stahl's thought. For example, in his *Fundamenta Chemiae* (1723), which presented his earliest views, he stated that gold and silver were comprised of Becher's three earths (vitriolic, inflammable, and mercurial). Later, these became "ash" or "primitive earth," phlogiston, and "mercurial spirit." See Stahl, *Philosophical Principles,* 15.

75. On the German case, see Oldroyd, "Some Phlogistic Mineralogical Schemes." For the French case, see Homberg, "Observation sur le fer au verre ardent"; Geoffroy, "Experiences sur les métaux." Geoffroy, for example, stated that metals can be resolved into "soufre ou une substance huileuse" and "chaux," a form of earth. (212).

76. Boerhaave presented an in-depth discussion of the metals in *Elements of Chemistry,* 1: 19–27.

77. Ibid., 1:380. One of J. J. Becher's three earths was "vitrifiable earth," which was also the matter that remained after a metal was calcined; see Partington, *History of Chemistry,* 2:644–45.

78. Boerhaave, *Elements of Chemistry,* 1:381–82. Boerhaave's experiments are discussed fully in the next chapter.

79. Ibid., 1:379, 381–82, quotation at 381–82.

80. Ibid., 1:23, 26–27.

81. Holmes, *Eighteenth-Century Chemistry,* 45–49; Partington, *History of Chemistry,* 2:659–86.

82. See Duncan, *Laws and Order*; Kim, *Affinity, That Elusive Dream*; Holmes, *Eighteenth-Century Chemistry*; Taylor, "Making Out a Common Disciplinary Ground."

83. Duncan, "Some Theoretical Aspects," 189–94; Duncan, *Laws and Order,* 115, 182–87.

84. Geoffroy, "Table des différents rapports," quotation at 257: "Toutes les fois que deux substances qui ont quelque disposition à se joindre l'une avec l'autre, se trouvent unies ensemble; s'il en survient une troisième qui ait plus de rapport avec l'une des deux, elle s'y unit en faisant lâcher prise à autre."

85. Ibid. On Geoffroy's choice of "rapport," see Duncan, "Some Theoretical Aspects," 182, 184; Holmes, *Eighteenth-Century Chemistry,* 38–39.

86. Geoffroy, "Table des différents rapports," 257–58, quotation at 258: "les Chymistes y trouveront une méthode aisée pour découvrir ce qui se passe dans plusieurs de leurs opérations difficiles à démêler, & ce qui doit résulter des mêlanges qu'ils font de différents corps mixtes."

87. Holmes, "Communal Context"; Klein, "E. F. Geoffroy's Table."

88. Boerhaave, *Elements of Chemistry,* 1:386 (quotation), 390–91, 404.

89. Ibid., 1:398, 397.

90. Ibid., 1:391, 413–15.

91. Ibid., 1:420–21.

92. Geoffroy, "Table des différents rapports," 269; Geoffroy, "Eclaircissements sur la table"; Duncan, *Laws and Order,* 93–96, 117–19.

93. Boerhaave, *Elements of Chemistry,* 1:454–55.

94. Ibid., 1:468–69. Spirit of niter was concentrated nitric acid; spirit of salt was hydrochloric acid.

95. Geoffroy did try to incorporate aqua regia in the table. As he explained in the text of the *mémoire,* Geoffroy represented the ability of aqua regia to dissolve gold by placing gold at the bottom of the series of displacements for "acid of sea salt," separated from the other substances in the series by two blank spaces. This move, however, undermined Geoffroy's contention to have his table readable "at a glance," because one would have to read the accompanying text to understand why the gold was there. See Geoffroy, "Table des différents rapports," 265–66.

96. Duncan, *Laws and Order,* 94–106.

97. Eklund, "Chemical Analysis and the Phlogiston Theory," 121–33; Lehman, "Mid-Eighteenth-Century Chemistry."

98. Anderson, "Boerhaave to Black"; Donovan, *Philosophical Chemistry*; Christie, "William Cullen and the Practice of Chemistry"; Wightman, "William Cullen and the Teaching of Chemistry," pts. 1 and 2; Clow, "Herman Boerhaave and Scottish Chemistry." The similarities between Boerhaave and Cullen (and Black) tended to be underemphasized by Donovan because he did not analyze and compare the pedagogical structure of courses.

99. On Rouelle, see Rappaport, "Rouelle and Stahl"; Lehman, "Mid-Eighteenth-Century Chemistry"; Duncan, *Laws and Order*, 112, 165. On the Scottish chemists, see Taylor, "Unification Achieved"; Taylor, "Making Out a Common Disciplinary Ground"; Donovan, *Philosophical Chemistry*, 129–43.

CHAPTER SEVEN: FROM ALCHEMY TO CHEMISTRY

1. Louis, Chevalier de Jaucourt, "Voorhout," in Diederot and d'Alembert, *Encyclopédie* (1764), 15:471, as cited in Lindeboom, *Herman Boerhaave*, 260.

2. Dobbs, *Foundations*; Debus, "Alchemy in an Age of Reason"; Principe, "Wilhelm Homberg"; Joly, "Quarrels between Étienne-François Geoffroy and Louis Lémery"; Kuhn, "Robert Boyle." Dobbs argued that Boerhaave was an unabashed believer in transmutation, a view critiqued in Scopa, "Boerhaave on Alchemy."

3. Principe, "Revolution Nobody Noticed?"; Powers, "'*Ars sine arte.*'" For the long-term view of alchemy from the eighteenth century, see Principe and Newman, "Some Problems."

4. Boerhaave, *Elements of Chemistry*, 1:19–27.

5. "Voorstel van Boerhaave aan C. en B. betr. verbeteringen in het Chemisch Laboratorium," August 1718; Res. Cur., 8 August 1718; and "Inventaris van het Laboratorium Chymicum," 8 November 1718; in Molhuysen, *Bronnen*, 4:150*, 302–33, and 151*–28, respectively.

6. See "Boerhaave's Chemical Education," in chap. 2. Also, Boerhaave, *Elements of Chemistry*, 1:26–27. On Boyle, see Principe, *Aspiring Adept*, 43–46; also Van Helmont, "Progymnasma Meteori," in *Ortus Medicinae*, 70–71, secs. 14–18; Homberg, "Essais de chimie," 46.

7. See Newman, *Summa Perfectionis*; Newman, *Gehennical Fire*, 95–98.

8. "Collegium Chemicum," in VMA, MS 3, 20v.

9. Boerhaave, *Elements of Chemistry*, 1:26–27.

10. Principe, *Aspiring Adept*, 43–46.

11. Boerhaave, *Elements of Chemistry*, 1:26.

12. See, for example, Becher, *Chemischer Glücks-Hafen*, 88–96; Smith, *Business of Alchemy*, 174–76.

13. VMA, MS 5, 16v–17r, 10v–11r.

14. For the roots of this theory, see Newman, *Summa Perfectionis*, 143–67; Newman, *Gehennical Fire*, 95–98.

15. On Isaac Holland, see Van Spronsen, "Beginning of Chemistry," 331; Zacher, "Die Bedeutung der Holländer."

16. Boerhaave, "De Mercurio Experimenta," pt. 2, 343–59, at 354. I have been unable to locate the text of Isaac Hollandus from which this recipe derived.

17. VMA, MS 5, 43r–v. Boerhaave later performed variations on this process; see 123r–v, 152v, and the section titled "Mercury Experiments" in this chapter.

18. Philalethes stated that the mercury obtained in this way was a "Mercury Crude, and no way fit for Medicines [i.e., alchemical operations]." See Newman, *Gehennical Fire*, 323n66.

19. VMA, MS 5, 123r–24r, 149v, 151v, 152v. See also Boerhaave, "De Mercurio Experimenta," pt. 2, 353–55.

20. Principe, "Wilhelm Homberg"; Homberg, "Suite de l'article des essais de chymie." On perpetual mines, see Becher, *Experimentum Novum Curiosum*.

21. "Elemens de chymie," in VMA, MS 130, 2r–108v.

22. This experiment is described in detail in the next section.

23. VMA, MS 5, 67r: "Quid est de hoc pulvere? inquire per nova experimenta. imprimis cohoba [mercurium] cum suo proprio hoc pulvere." "Cohobation" is the process of distilling a sample repeatedly by collecting the distillate after each run and returning it to the cucurbit (i.e., still pot) for further distillation.

24. Ibid., 118r–19r.

25. Ibid., 119v: "ut ita scirem, si quid fixi metallici inesset?"

26. On Suchten's process, see Newman and Principe, *Alchemy Tried in the Fire*, 50–58. On Suchten generally, see Hubicki, "Alexander von Suchten." On Suchten and Philalethes, see Newman, "Prophecy and Alchemy"; Newman, *Gehennical Fire*, 135–41.

27. VMA, MS 5, 140r–41r. For a full explanation of Philalethes's process, see Newman, *Gehennical Fire*, 92–169.

28. VMA, MS 5, 143v.

29. Ibid., 171r.

30. Ibid., 184v–85v, 191r, 191v.

31. Boerhaave, *Elements of Chemistry*, 1:72.

32. Two prominent examples of this literature were Geoffroy, "Des supercheries concernent la Pierre Philosophale"; Stahl, *Bedenken von der Gold-Macherey*.

33. Boerhaave, *Elements of Chemistry*, 1:72.

34. Ibid., 1:75. For Boerhaave's claims in his oration, see chap. 4, and Boerhaave, *Sermo Academicus de Chemia*; Boerhaave, "Discourse on Chemistry Purging Itself of Its Own Errors," in *Boerhaave's Orations*, 180–213.

35. Boerhaave, *Elements of Chemistry*, 1:76, 77.

36. Boerhaave to Mortimer, 12 July 1733, in Lindeboom, *Boerhaave's Correspondence*, 1:206–9. Quotation (my translation) at 127: "de miris dotibus argenti vivi per laboriosissima experimenta exploranti." On Boerhaave's election as a fellow of the Royal Society, see Lindeboom, *Herman Boerhaave*, 170–71; Underwood, *Boerhaave's Men at Leyden*, 55, 133–34, 152, 154. On Boerhaave's illness, see Lindeboom, *Herman Boerhaave*, 186–87.

37. Boerhaave to Mortimer, 18 February 1734, and 3 March 1737, in Lindeboom, *Boerhaave's Correspondence*, 208–9 and 214–15, respectively; Boerhaave, "De Mercurio Experimenta."

38. Boerhaave, "De Mercurio Experimenta," pt. 1, 145: "Nulli profexto mortalium indagini rerum naturalium acrem adeo, & pertinacem, operam dedere, aut versande per varios explorandi modos materiae improbum adeo laborem adhibuerunt, quam Alchemistae."

39. Ibid., 146: "deliri, fabulosi, mendaces, & vani explodantur. Enimvero verbis severi, divites promissis, rem interea ipsam atra adeo caligine condunt, ut arcana revelata nolle videantur. Sapientum ideo plurimi judicant, impossibile protsus naturae, & arti, quod promittunt."

40. Ibid.: "contigit mihi chemica scrutanti, & evolventi Alchemistarum scripta, videre, unum mentem omnibus."

41. Ibid., 146–48.

42. Ibid., 148: "Si vero arte difficillima repurgatur penitus ab ea peregtina macula, tum demum haberi liquidum; metallicum; pondersoissimum; simplicissimum; nulla arte, nec natura, unquam in diversa divisibile; in quo soiluti cujusque metalli semen vivicatum, se perfectissime multiplicaret; in quo aurum ipsum deliquescens, fotum, maturatum, soret laboris supremum pretium, quaesitum adeo, adeo decantatum."

43. In the *Elementa*, Boerhaave combined the *operatio* and *effectus* together under the *experimentum*.

44. Boerhaave, "De Mercurio Experimenta," pt. 1, 149–50, at 150: "Vitrum sic paratum curavi affigendum ad caudicem tundentem molendinae fulloniae, noctes, diesque, mobilis, modo spiraret ventus; . . . fuit semper ad perpendiculum elevatu, & demissu, concissum." Boerhaave's original experimental notes on this experiment may be found in VMA, MS 5, 136r. A "fuller's hammer" was

employed by blacksmiths and in metallurgy for spreading and beating metals into plates. Boerhaave was obviously describing a mechanical hammering device, probably powered by a waterwheel along one of Leiden's rivers or canals. He did not record in the notes where this device was located or to whom it belonged.

45. Boerhaave, "De Mercurio Experimenta," pt. 1, 150.

46. Boerhaave, *Elements of Chemistry*, 1:381–82.

47. VMA, MS 5, 118r–19r.

48. Boerhaave, "De Mercurio Experimenta," pt. 1, 151–60.

49. See the section in this chapter titled "Alchemical Work"; Principe, "Wilhelm Homberg."

50. Boerhaave, "De Mercurio Experimenta," pt. 1, 160–61.

51. Ibid., 162: "Ignis igitur, ex his experimentis, non demonstratur Philosophorum Sulfur fixans Mercurium in metalla."

52. Ibid., 163: "Tuti a fallacibus scriptis & praescriptis, Sophistarum, qui ex Mercurio, & igne talia promittunt intra breve tempus, paucosue menses: sane intra plures annos ne inchoamenti quidem primi vel minima indica."

53. Ibid., 165: "Scripsit Geber, Mercurium purum auro ponderosiorem." On Geber, see Newman, *Summa Perfectionis*, 162–67.

54. Newman, *Summa Perfectionis*, 163–64.

55. Boerhaave, "De Mercurio Experimenta," pt. 1, 165–66.

56. Ibid., 166: "1. Si Mercurius defoecatus levior fit: tum defoecatus redditur per aurum & plumbum. Arte Suchetnii, & Philalethae, manet idem."

57. Ibid. : "5. An Mercurius deponit gravissimam partem sui in auro? An haec deposita est semen auri?"

58. Newman, *Gehennical Fire*, 165–68.

59. Boerhaave, "De Mercurio Experimenta," pt. 1, 166: "An Mercurius, opere continuato, tandem posset densari in pondus auri? . . . Examinent judices idonei."

60. Ibid., pt. 2, 344–48. See also VMA, MS 5, 153r, 179r.

61. Boerhaave, *Elements of Chemistry*, 1:460–62. On Boerhaave's earlier experiments with "resuscitating" alkali, see VMA, MS 1, 21r, and MS 5, 43r–v.

62. Boerhaave, "De Mercurio Experimenta," pt. 2, 348–53. The notes for the published experiments are found in VMA, MS 5, 131r, 146v.

63. Boerhaave, "De Mercurio Experimenta," pt. 3, 369: "Cruda coquit, perficit vilia, metalla, alia quaelibet corpora prompte attenuat, resolvit, in humidum convertit radicale; sic, ad arcana medicinae, as secreta hermeticae artis, princips instrumentum jure laudatur, quo quid, summo."

64. Ibid., 370.

65. Ibid., 371: "Atque Aurem, Argentumque, pura, fixa, ipsique Mercurio sincero metalla quam similla; sive originem spectaveris, sive materiem. Sequi his statuunt, si permiscetur perfecto Mercurius metallo, atque iterum igne inde expellitur intra bene clausa vasa vitrea, partem metalli puri in se tracturam Mercurium, simulque sordidum Mercurii secreturam puro."

66. On the (early) gold experiments, see VMA, MS 5, 18v–20r, 77r–82r, 99r–100v; on silver, see 20v–22r, 64v–70v, 173r, 178r. The specific amalgamation and distillation experiments on gold are found at 18v–19r and 78r–80v, respectively.

67. Ibid., 181v–82v.

68. See ibid., 183r. Compare this with Boerhaave, "De Mercurio Experimenta," pt. 3, 371–74. The experimental notes only recorded the first 250 distillations, when Boerhaave paused after each set of 50 distillations to weight his products. The record of the remaining 627 distillations are lost and are only mentioned in the published paper.

69. Boerhaave, "De Mercurio Experimenta," pt. 3, 374: "ut molimine tanto densitatem haud mutaterit Mercurius, neque ulla parte sui leviore fuerit liberatus."

70. Ibid., 375–76: "Non firmatur hinc opinio, quae narrat, ignem metallis, vel mercurio, con-

crescere posse in augmentum, vel generationem, alicuius metallici; aut mutationem ipsius metalli stabilem."

71. Boerhaave to Bassand, 3 October 1732, in Lindeboom, *Boerhaave's Correspondence*, 310–11: "primo humorem pinguem, spissum, adipis instar, coloris vero ut plurimum ex flavo viridescente, quem Germano fossores proprio vocabulo Guhr appellant." See also Boerhaave's rejection of Boyle and Homberg, in Boerhaave to Bassand, 20 July 1725, ibid., 336–37. On the origin of Guhr-like fluid theories of metallogenesis, see Norris, "Early Theories." For a later example see Newman, "Newtons's Theory of Metallic Generation."

CONCLUSION: BOERHAAVE'S LEGACY

1. "Memorie tot het rapport van den Secretaris van Royen . . . geweest d'heer Boerhaven te spreken over de capaciteit van die geenen, welke hem zoude kunnen sucederen," 22 April 1738, in Molhuysen, *Bronnen*, 5:5:62*–65*, at 64*. See also Lindeboom, *Herman Boerhaave*, 224–27.

2. On physico-mathematics and constructive geometry, see Dear, *Discipline and Experience*; Garrison, "Newton."

3. Broman, *Transformation*, 16–17; Lindeboom, *Herman Boerhaave*. Most of the evidence on medical students and other visitors stopping in Leiden is anecdotal. See, for example, the account of Albrecht von Haller: Von Haller, *Albrecht Hallers Tagebücher*.

4. Frijhoff, *Société néerlandaise*, 103–7.

5. For Boerhaave as a "Hippocratic" physician (and chemist), see chap. 1, and Knoeff, "Practicing Chemistry." On the Hippocratic revival, see Nutton, "Hippocrates in the Renaissance"; Lonie, "'Paris Hippocrates.'"

6. Albinus and Boerhaave published a revised edition of Vesalius's *Opera Omnia*. See Vesalius, *Opera Omnia Anatomica et Chirurgica*.

7. Von Haller, *Albrecht Hallers Tagebücher*, 98–99, 101; Lindeboom, *Herman Boerhaave*, 122; Lindeboom, "Oosterdijk Schacht, Herman," and "Albinus, Bernard Sigfried," in *Dutch Medical Biography*, cols. 1471–72 and 9–12, respectively.

8. Cf. Lindeboom, *Herman Boerhaave*, 167–68; Res. Cur., 26 February and 12 March 1729, 9 January 1731, in Molhuysen, *Bronnen*, 5:69–70, 95. For Boerhaave's list for the chair of chemistry, see Molhuysen, *Bronnen*, 5:28*.

9. The publication of Boerhaave's *Elementa* and "De Mercurio Experimenta" are discussed in chapters 6 and 7, respectively. On the editions of the works of Aretaeus of Cappadocia (a Greek empiricist physician from the first century) and Jan Swammerdam, see Lindeboom, *Herman Boerhaave*, 184–86, 198–201.

10. Lindeboom, *Herman Boerhaave*, 224–27; Res. Cur., 8 May and 8 August 1738, in Molhuysen, *Bronnen*, 5:185, 186.

11. Cf. Lindeboom, *Herman Boerhaave*, 360–74.

12. Cf. Risse, "Clinical Instruction"; Lindeboom, *Herman Boerhaave*, 360–62. Harm Beukers has presented a more skeptical view of Boerhaave's influence on clinical instruction; see Beukers, "Clinical Teaching."

13. Broman, *Transformation*, 13–14, 76–80. Von Haller had first been exposed to Boerhaave's *Institutiones Medicae* while he was a medical student at Tübingen; cf. Lindeboom, *Herman Boerhaave*, 75.

14. Lindeboom, *Herman Boerhaave*, 363–64; Lesky and Wandruszka, *Gerard van Swieten*; Van Swieten, *Commentaria*. Van Sweiten's *Commentaria* was enormously popular and influential, appearing in over forty editions in the eighteenth century. See Lindeboom, *Bibliographia Boerhaaviana*, 41–47.

15. Lindeboom, *Herman Boerhaave*, 364, 74–75. On the rivalry between Von Haller and Van Swieten over Boerhaave's legacy, see Lesky, "Albrecht von Haller."

16. Lindeboom, *Herman Boerhaave*, 366–67.

17. Emerson, "Founding of the Edinburgh Medical School"; Cunningham, "Medicine to Calm the Mind"; Guthrie, "Influence of the Leyden School"; Comrie, "Boerhaave and the Early Medical School at Edinburgh."

18. Sigerist, "Boerhaave's Influence on American Medicine."

19. Cf. Eddy, *Language of Mineralogy*; Anderson, "Boerhaave to Black"; Donovan, *Philosophical Chemistry*.

20. Cf. Meinel, "*Artibus Academicus Inserenda*."

21. See Peter Shaw's English translation of the *Elementa*: Boerhaave, *New Method of Chemistry*.

22. See, for example, Boerhaave, *Elements of Chemistry*, 1:52.

23. I discuss this in chapters 5 and 6. For Boerhaave's encouragement to his students, see, for example, Boerhaave, *Elements of Chemistry*, 1:246–47.

24. Shaw, *Chemical Lectures*, 25–28. These experiments are an abbreviated version of those in the *Elementa* at the start of Boerhaave's discussion of fire; cf. Boerhaave, *Elements of Chemistry*, 1:78–107. On Shaw's lectures and program, see Gibbs, "Peter Shaw"; Golinski, "Peter Shaw."

25. Wightman, "William Cullen and the Teaching of Chemistry (I) and (II)"; Donovan, *Philosophical Chemistry*; Anderson, "Boerhaave to Black." Both Cullen and Black began their discussion of heat with a discussion of expansion, and each demonstrated a version of Boerhaave's expanding bar experiments; cf. Black, *Lectures*, 40–41; Donovan, *Philosophical Chemistry*. For Cullen's thermometer experiments, see, for example, Cullen, "Cold Produced by Evaporating Fluids."

26. Fontenelle, "Eloge de M. Boerhaave." For a general account of Boerhaave's "influence" on other, prominent chemists, see Greenaway, "Boerhaave's Influence."

27. Macquer, *Dictionnaire*, xxvi–xxvii.

28. On the instrument theory in lectures at the Jardin, see Lehman, "Mid-Eighteenth-Century Chemistry"; Rappaport, "G.-F. Rouelle."

29. Klein, "Blending Technical Innovation," esp. 150–55; Klein and Lefèvre, *Materials in Eighteenth-Century Science*.

30. Cf. Klein, "Experimental History"; Klein, "Blending Technical Innovation."

31. Boerhaave, *Elements of Chemistry*, 1:52–78.

32. Ibid., vol. 2; Boerhaave, *Libellus de Materie Medica*.

33. Sumner, "Michael Combrune"; Fors, "Occult Traditions and Enlightened Science."

34. Cf. Principe, *New Narratives*; Eddy, *Language of Mineralogy*; Klein and Lefèvre, *Materials in Eighteenth-Century Science*; Kim, *Affinity, That Elusive Dream;* Principe, *Aspiring Adept*; and Newman and Principe, *Alchemy Tried in the Fire*. Many of the recent studies dedicated to reconceiving eighteenth-century chemistry have been inspired by the foundational work of Frederic Holmes; see Holmes, *Eighteenth-Century Chemistry*.

35. Guerlac, *Lavoisier*, 20–22, 25–28, 120–22. See also Melhado, "Oxygen, Phlogiston and Caloric"; Melhado, "Chemistry, Physics, and the Chemical Revolution." Guerlac asserts that Boerhaave saw fire and air as *physical* agents and, as such, was hesitant to give them a role in the chemical composition of bodies. In regard to air, I have shown in chaps. 5 and 6 that this was not the case.

36. Lavoisier, *Elements*, 1. For Lavoisier's discussion of Boerhaave's work on air, see Lavoisier, *Opuscules*, 26–30.

37. Shaw, *Chemical Lectures*, 1.

Bibliography

ARCHIVES / MANUSCRIPTS

British Library, Sloane Collection, London, UK. MSS 1216, 1268, 1277, 1278, 1284–88, 1291, 1292, 2793, 2969, 3179, and 3860.

Fundamental'naya Biblioteka Voenno-Meditsinskoi Akademii (VMA), St. Petersburg, Russia. Fund XIII. MSS 1–9, 19, 25, 26, 28, 30, 34, 36, 41, 43, 44, 105, 128, 130, 131, 137, and 150.

Royal College of Surgeons Archives, London, UK. MS 42.b.49.

Leiden University Archives (Universiteitsbibliotheek, Rijksuniversiteit Leiden), Netherlands. MS BPL 772.

PUBLISHED SOURCES

Agricola, Georgius. *De Re Metallica*. Edited and translated by Herbert Clark Hoover and Lou Henry Hoover. New York: Dover, 1950.

Ahonen, Kathleen. "Johann Rudolph Glauber: A Study of Animism in Seventeenth-Century Chemistry." Ph.D. dissertation, University of Michigan, 1972.

Alembert, Jean le Rond d'. *Preliminary Discourse to the Encyclopedia of Diderot*. New York: Bobbs-Merrill, 1963.

Alpers, Svetlana. *The Art of Describing: Dutch Art in the Seventeenth Century*. Chicago: University of Chicago Press, 1983.

Anderson, Robert G. W. "Boerhaave to Black: The Evolution of Chemistry Teaching." *Ambix* 53 (2006): 237–54.

Anon. "*Philosophia Naturalis*, Auctore Wolferdo Senguerdo, Philosophiae Professore." *Acta Eruditorum* (March 1682): 83–84.

Atkins, Edward R. "Samuel Johnson's 'Life of Boerhaave.'" *Journal of Chemical Education* 19 (1942): 103–6.

Bacon, Francis. *The Advancement of Learning*. Edited by G. W. Kitchin. London: J. M. Dent, 1973.

———. *The New Organon*. Edited by Fulton H. Anderson. New York: Macmillan, 1960.

Baumann, E. D. *François Dele Boë Sylvius*. Leiden: Brill, 1949.

Becher, Johann Joachim. *Chemischer Glücks-Hafen, Oder Grosse Chymische Concordantz und Collection von funffzehen hundert Chymischen Processen*. Frankfurt: J. G. Schiele, 1682.

———. *Experimentum Novum Curiosum de Minera Arenaria Perpetua sive Prodromus Historiae*. Frankfurt: Martinus Georgius Weidmannus, 1680.

Beguin, Jean. *Tyrocinium Chymicum e Naturae et Manuali Experientia Deppromptum, Sexta Editione, Studio & Opera Christophori Gluckradi*. Berger: Boreck, 1625.

Bennet, J. A. "The Mechanics' Philosophy and the Mechanical Philosophy." *History of Science* 24 (1986): 1–28.

———. "Robert Hooke as Mechanic and Natural Philosopher." *Notes and Records of the Royal Society of London* 35 (1980): 33–48.

Bensaude-Vincent, Bernadette, and Christine Blondel, eds. *Science and Spectacle in the European Enlightenment*. Aldershot, UK: Ashgate, 2008.

Bensaude-Vincent, Bernadette, and Christine Lehman. "Public Lectures of Chemistry in Mid-Eighteenth-Century France." In *New Narratives in Eighteenth-Century Chemistry*, edited by Lawrence M. Principe, 77–96. Dordrecht: Springer, 2007.

Beretta, Marco. "The Historiography of Chemistry in the Eighteenth Century: A Preliminary Survey and Bibliography." *Ambix* 39 (1992): 1–10.

Berkvens-Stevelinck, C., J. Israel, and G. H. M. Posthumus Meyjes, eds. *The Emergence of Toleration in the Dutch Republic*. Studies in the History of Christian Thought 76. Leiden: Brill, 1997.

Bertoloni Meli, Domenico. "The Collaboration between Anatomists and Mathematicians in the Mid-Seventeenth Century with a Study of Images as Experiments and Galileo's Role in Steno's *Myology*." *Early Science and Medicine* 13 (2008): 665–709.

Beukers, Harm. "Acid Spirits and Alkaline Salts: The Iatrochemistry of Franciscus dele Boë, Sylvius." *Sartoniana* 12 (1999): 39–58.

———. "Clinical Teaching at Leiden from Its Beginning until the End of the Eighteenth Century." *Clio Medica* 21 (1987–88): 139–52.

———. "Het Laboratorium van Sylvius." *Tijdschrift voor Geschiedenis derGeneeskunde, Natuurwetenschappen, Wiskunde, en Techniek* 3 (1980): 28–36.

———. "Mechanische Principes bij Franciscus Dele Boë Sylvius." *Tijdschrift voor Geschiedenis der Geneeskunde, Natuurwetenschappen, Wiskunde, en Techniek* 5 (1982): 6–15.

Black, Joseph. *Lectures on the Elements of Chemistry*. 2 vols. Edited by John Robison. Edinburgh: Mundell, 1803.

Boerhaave, Herman. *Andreae Vesalii Opera Omnia Anatomica et Chirurgica*. 2 vols. Edited by Herman Boerhaave and Bernard Siegfried Albinus. Leiden, 1725.

———. *Aphorismi de Cognoscendis et Curandis Morbis*. Leiden: Apud Johannen vander Linden, 1709.

———. *Boerhaave's Aphorisms concerning the Knowledge and Cure of Diseases*. Edited and translated by J. Delacoste. London: R. Crowse and W. Innys, 1715.

———. *Boerhaave's Correspondence*. 3 vols. Edited by G. A. Lindeboom. Leiden: Brill, 1962–70.

———. *Boerhaave's Orations: Translated, with Introductions and Notes*. Edited by E. Kegel-Brinksgreve and A. M. Luyendijk-Elshout. Leiden: Brill, 1983.

———. "Commentariolus de Familia, Studiis, Vitae Cursu, &c." In *Herman Boerhaave: The Man and His Work*, by G. A. Lindeboom, 375–86. London: Methuen, 1968.

———. *De Distinctione Mentis a Corpore*. Leiden: Abraham Elzevier, 1690.

———. "De Mercurio Experimenta." *Philosophical Transactions of the Royal Society of London*, pt. 1, 38, no. 430 (1733): 145–76; pt. 2, 39, no. 443 (1736): 343–59; pt. 3, no. 444 (1736): 368–76.

———. *De Usu Ratiocinii Mechanici in Medicina*. Leiden, 1703.

———. "De Utilitate Explorandorum in Aegris Excrementorum ut Signorum." Thesis. Harderwyk, 1693.

———. *Dr. Boerhaave's Academical Lectures on the Theory of Physick*. 3rd ed. corrected. London: For J. Rivington et al., 1766.

———. *Elementa Chemiae, Quae Anniversario Labore Docuit, in Publicis, Privatisque Scholis Hermannus Boerhaave*. 2 vols. Leiden: Apud Isaacum Severnium, 1732.

———. *Elements of Chemistry, being the Annual Lectures of Herman Boerhaave*. 2 vols. Translated by Timothy Dallowe. London: J. and J. Pemberton, 1735.

———. *Index alter Plantarum Quae in Academico Lugduno-Batavo Aluntur*. 2 vols. Leiden: Sumptibus Auctoris, 1720.

———. *Index Plantarum, Quae in Horto Academico Lugduno-Batavo Reperiuntur*. Leiden: Apud Cornelius Boutestein, 1710.

[———]. *Institutiones et Experimenta Chemiae*. 2 vols. Paris, 1724.

———. *Institutiones Medicae; in Usus Annuae Exercitationis Domesticos*. Leiden: Apud Johannen vander Linden, 1708.

———. *Libellus de Materie Medica et Remediorum Formulis, Quae Serviunt Aphorismis de Cognoscendis et Curandis Morbis*. Leiden: Apud Isaacum Severnium, 1719.

———. *A Method of Studying Physick*. 8 vols. Translated by Mr. Samber. London: C. Rivington, B. Creake, and J. Sachfield, 1719.

[———]. *A New Method of Chemistry; including the History and Practice of the Art*. 2 vols. 3rd ed. Translated by Peter Shaw. London: T. Longman, 1741.

[———]. *A New Method of Chemistry; including the Theory and Practice of the Art*. Translated by Peter Shaw and Ephriam Chambers. London: J. Osborn and T. Longman, 1727.

———. *Oratio de Commendando Studio Hippocratico*. Leiden, 1701.

———. *Oratio qua Repurgatae Medicinae Facilis Asseritur Simplicitas*. Leiden, 1709.

———. *Sermo Academicus de Chemia suos Errores Expurgante*. Leiden: Sumptibus Petri vander Aa, 1718.

———. *Sermo Academicus de Comparando Certo in Physicis*. Leiden: Sumptibus Petri vander Aa, 1715.

———. *Sermo Academicus quem Habuit quum Honesta Missione Impetatra Bontanicum et Chemicum Professionem Publice Poneret XXVII Aprilis 1729*. Leiden, 1729.

Bohn, Johannes. *Dissertationes Chymico-Physicae, Chemiae Finem, Instrumenta & Operationes Frequentiores Explicantes, cum Indice Rerum & Verborum*. Lipsae: Joh. Fredericus Gleditschius, Literis Christianus Gözl, 1685.

Bohrod, Milton G. "Medical Genealogy." *Journal of the History of Medicine* 24 (1969): 292–93.

Bougard, Michel. *La chimie du Nicolas Lemery*. Turnhout: Brepols, 1999.

Bos, E. P., and H. A. Krop, eds. *Franco Burgersdijk (1590–1635): Neo-Aristotelianism in Leiden*. Amsterdam: Rodopi, 1993.

Boyle, Robert. *Experiments, Notes, &tc. About the Mechanical Origine or Production of Divers Particular Qualities*. London, 1675.

———. "Of the Incalescence of Quicksilver with Gold." *Philosophical Transactions of the Royal Society of London* 10 (1676): 515–33.

———. *The Producibleness of Chymical Principles*. Oxford, 1680.

———. *The Sceptical Chymist*. London, 1661; London: Dent, 1967.

———. *Some Considerations Touching the Usefulness of Experimental Natural Philosophy*. Oxford, 1663.

Briggs, Robin. "The Académie Royale des Sciences and the Pursuit of Utility." *Past and Present* 131 (May 1991): 38–88.

Brock, William. *The Norton History of Chemistry*. New York: Norton, 1992.

Brockliss, L. W. B. *French Higher Education in the Seventeenth and Eighteenth Centuries: A Cultural History*. Oxford: Clarendon Press, 1987.

———. "Medical Teaching at the University of Paris, 1600–1720." *Annals of Science* 35 (1978): 221–52.

Brockliss, Laurence, and Colin Jones. *The Medical World of Early Modern France*. Oxford: Oxford University Press, 1997.

Broman, Thomas H. *The Transformation of German Academic Medicine*. Cambridge: Cambridge University Press, 1996.

Brown, Theodore M. *The Mechanical Philosophy and the "Animal Economy."* New York: Arno, 1981.

———. "Medicine in the Shadow of the *Principia*." *Journal of the History of Ideas* 48 (1987): 629–48.

Burnet, Pierre. *Les physiciens hollandais et la méthode expérimantale en France au XVIIIe siècle*. Paris: Alcan, 1931.

Burton, William. *An Account of the Life and Writings of Herman Boerhaave*. 2nd ed. London: Henry Lintot, 1743.

Büttner, Johannes. "Die physikalische und chemische Untersuchung von Blut im 17 und 18 Jahrhundert: Zur Bedeutung von Robert Boyle's 'Memoirs for the Natural History of Human Blood' (1684)." *Medizin Historisches Journal* 22 (1987): 185–96.

Bylebyl, Jerome J. "The School of Padua: Humanistic Medicine in the Sixteenth Century." In *Health, Medicine, and Mortality in the Sixteenth Century*, edited by Charles Webster, 335–70. Cambridge: Cambridge University Press, 1979.

Bynum, William F. "The Anatomical Method, Natural Theology and the Function of the Brain." *Isis* 64 (1973): 445–68.

Bynum, William F., and Roy Porter, eds. *Medical Fringe and Medical Orthodoxy, 1750–1850*. London: Croom Helm, 1987.

Cantor, David, ed. *Reinventing Hippocrates*. Aldershot, UK: Ashgate, 2001.

Chang, Hasok. *Inventing Temperature: Measurement and Scientific Progress*. New York: Oxford University Press, 2004.

Chang, Ku-Ming (Kevin). "Fermentation, Phlogiston and Matter Theory: Chemistry and Natural Philosophy in Georg Ernst Stahl's *Zymotechnia Fundamentalis*." *Early Science and Medicine* 7 (2002): 31–64.

———. "From Oral Disputation to Written Text: The Transformation of the Dissertation in Early Modern Europe." *History of Universities* 19 (2004): 129–87.

Christie, J. R. R.. "Historiography of Chemistry in the Eighteenth Century: Herman Boerhaave and William Cullen." *Ambix* 41 (1994): 4–19.

———. "William Cullen and the Practice of Chemistry." In *William Cullen and the Eighteenth Century Medical World*, edited by A. Doig, J. P. S. Ferguson, I. A. Milne, and R. Passmore, 98–109. Edinburgh: Edinburgh University Press, 1993.

Christie, J. R. R., and J. V. Golinski. "The Spreading of the Word: New Directions in the Historiography of Chemistry, 1600–1800." *History of Science* 20 (1982): 235–66.

Clericuzio, Antonio. "Teaching Chemistry and Chemical Textbooks in France: From Beguin to Lemery." *Science and Education*. 15 (2006): 335–55.

Clow, Archibald. "Herman Boerhaave and Scottish Chemistry." In *An Eighteenth-Century Lectureship in Chemistry*, edited by Andrew Kent, 41–48. Glasgow: University of Glasgow, 1950.

Cohen, Ernst. "Herman Boerhaave und seine Bedeutung für die Chimie." *Janus* 23 (1912): 221–78.

Cohen, Ernst, and W. A. T. Cohen–de Meester. "Daniel Gabriel Fahrenheit." *Verhandelingender Nederlandsche Akademie van Wetenschappen, Amsterdam*, sec. 1, 16, no. 2 (1936): 1–37.

———. "Katalog der Wiedergefunden Manuskripte und Briefwechsel von Herman Boerhaave." *Verhandelingender Nederlandsche Akademie van Wetenschappen, Afdeeling Natuurkunde* 40, no. 2 (1941): 1–45.

Cohen, I. B. *Franklin and Newton: An Inquiry into Speculative Newtonian Science*. Philadelphia: APS, 1956.

Coleby, L. J. M. "John Francis Vigani." *Annals of Science* 8 (1952): 46–60.

Collins, H. M. "Public Experiments and Displays of Virtuosity: The Core-Set Revisited." *Social Studies of Science* 18 (1988): 725–48.

Contant, Jean-Paul. *L'enseignement de la chymie au Jardin royale des plantes à Paris*. Cahors: Imprimerie Couleslant, 1952.

Comrie, J. R. "Boerhaave and the Early Medical School at Edinburgh." *Nederlandsch Tijdschrift voor Geneeskunde* 82 (1938): 4828–35.

Cook, Harold J. "Body and Passions: Materialism and the Early Modern State." *Osiris* 17 (2002): 25–48.

———. "Boerhaave and the Flight from Reason in Medicine." *Bulletin of the History of Medicine* 74 (2000): 221–40.

———. *Matters of Exchange: Commerce, Medicine, and Science in the Dutch Golden Age*. New Haven: Yale University Press, 2007.

———. "The New Philosophy in the Low Countries." In *The Scientific Revolution in National Context*, edited by Roy Porter and Mikulás Teich, 115–49. Cambridge: Cambridge University Press, 1992.

———. "Policing Health in London: The College of Physicians and the Early Stuart Monarchy." *Social History of Medicine* 2 (1989): 1–33.

———. "The Rose Case Reconsidered: Physic and the Law in Augustan England." *Journal of the History of Medicine* 45 (1990): 527–55.

———. *The Trials of an Ordinary Doctor: Joannes Groenevelt in Seventeenth-Century London*. Baltimore: Johns Hopkins University Press, 1994.

Crosland, Maurice P. *Historical Studies in the Language of Chemistry*. New York: Dover, 1962.

Cullen, William. "Of the Cold Produced by Evaporating Fluids, and Some Other Means of Producing Cold." In *Essays and Observations, Physical and Literary*, 2:159–71. 2nd ed. Edinburgh: John Balfour, 1770.

Cunningham, Andrew. "Medicine to Calm the Mind: Boerhaave's Medical System and Why It Was Adopted in Edinburgh." In *The Medical Enlightenment of the Eighteenth Century*, edited by Andrew Cunningham and Roger French, 40–66. Cambridge: Cambridge University Press, 1990.

———. "The Pen and the Sword: Recovering the Disciplinary Identity of Physiology and Anatomy before 1800. I. Old Physiology—The Pen." *History and Philosophy of the Biological and Biomedical Sciences* 33 (2002): 631–65.

———. "The Pen and the Sword: Recovering the Disciplinary Identity of Physiology and Anatomy before 1800. II. Old Anatomy—The Sword." *History and Philosophy of the Biological and Biomedical Sciences* 34 (2003): 51–76.

———. "Thomas Sydenham: Epidemics, Experiment, and the 'Good Old Cause.'" In *The Medical Revolution of the Seventeenth Century*, edited by Roger French and Andrew Wear, 164–90. Cambridge: Cambridge University Press, 1988.

Darnton, Robert. "Philosophers Trim the Tree of Knowledge: The Epistemological Strategy of the *Encyclopédie*." In *The Great Cat Massacre and Other Episodes in French Cultural History*. New York: Vintage, 1984.

Dear, Peter. *Discipline and Experience: The Mathematical Way in the Scientific Revolution*. Chicago: University of Chicago Press, 1995.

———. "Jesuit Mathematical Science and the Reconstruction of Experience in the Early Seventeenth Century." *Studies in History and Philosophy of Science* 18 (1987): 133–75.

———, ed. *The Literary Structure of Scientific Argument: Historical Studies*. Philadelphia: University of Pennsylvania Press, 1991.

———. "A Mechanical Microcosm: Bodily Passions, Good Manners and Cartesian Mechanism." In *Science Incarnate: Historical Embodiments of Natural Knowing*, edited by Christopher Lawrence and Steven Shapin, 51–82. Chicago: University of Chicago Press, 1998.

———. "Method and the Study of Nature." In *The Cambridge History of Seventeenth-Century Philosophy*, edited by Daniel Garber and Michael Ayers, 1:147–77. Cambridge: Cambridge University Press, 1998.

———. "Narratives, Anecdotes, and Experiments: Turning Experience into Science in the Seventeenth Century." In *The Literary Structure of Scientific Argument*, edited by P. Dear, 136–63. Philadelphia: University of Pennsylvania Press, 1991.

Debus, Allen G. "Alchemy in an Age of Reason: The Chemical Philosophers in Early Eighteenth-Century France." In *Hermeticism and the Renaissance: Intellectual History and the Occult in Early Modern Europe*, edited by Ingrid Merkel and Allen G. Debus, 231–50. Washington, DC: Folger Books, 1989.

———. *The Chemical Philosophy: Paracelsian Science and Medicine in the Sixteenth and Seventeenth Centuries*. 2 vols. New York: Science History Publications, 1977.

———. *Chemistry and Medical Debate: Van Helmont to Boerhaave*. Canton, MA: Science History Publications, 2001.

———. "Chemistry and Universities in the Seventeenth Century." *Estudios Acançados* 4 (1986): 173–96.

———. *The English Paracelsians*. New York: Franklin Watts, 1965.

———. "Fire Analysis and the Elements in the Sixteenth and Seventeenth Centuries." *Annals of Science* 23 (1967): 127–47.

———. *The French Paracelsians: The Chemical Challenge to Medical and Scientific Tradition in Early Modern France*. Cambridge: Cambridge University Press, 1991.

De Clercq, Peter. *At the Sign of the Oriental Lamp: The Musschenbroek Workshop in Leiden, 1660–1750*. Rotterdam: Erasmus, 1997.

De Maets, Carolus Ludovicus. *Prodromus Chemiae Rationalis, Ratiociniis Philosophicis, Observationibus, Medicis, &c Illustratae*. Leiden: Petrum de Graaf, 1684.

De Pater, C. "Experimental Physics." In *Leiden University in the Seventeenth Century*, edited by Lunsingh Scheurleer and Posthumus Meyjes, 309–27. Leiden: Universitaire Pers Leiden/Brill, 1975.

Diederot, Denis, and Jean d'Alembert, eds. *Encyclopédie, ou dictionnaire des sciences, des artes, et des métiers*. 35 vols. Paris, 1751–72.

Dijksterhuis, Fokko Jan. "Reading Up on the *Opticks*: Refashioning Newton's Theory of Light and Colors in Eighteenth-Century Textbooks." *Perspectives on Science* 16 (2008): 309–27.

Dobbs, Betty Jo Teeter. "From the Secrecy of Alchemy to the Openness of Chemistry." In *Solomon's House Revisited: The Organization and Institutionalization of Science*, edited by Tore Frängsmyr, 75–94. Canton, MA: Science History Publications, 1990.

———. *The Foundations of Newton's Alchemy, or the Hunting of the Greene Lyon*. New York: Cambridge University Press, 1975.

Donovan, Arthur. *Philosophical Chemistry in the Scottish Enlightenment: The Doctrines and Discoveries of William Cullen and Joseph Black*. Edinburgh: Edinburgh University Press, 1975.

Duclos, Samuel. "De l'augmentation du poids de certaine matieres par la calcination." In *Histoire des Académie royale des sciences, depuis 1666 jusqu'à 1699*, 1:34–37. Paris, 1777.

———. "Dissertation sur les principles des mixtes naturels." In *Histoire des Académie royale des sciences, depuis 1666 jusqu'à 1699*, 4:1–40. Paris, 1781.

Duncan, Alistair. *Laws and Order in Eighteenth-Century Chemistry*. Oxford: Clarendon Press, 1996.

———. "Some Theoretical Aspects of Eighteenth-Century Tables of Affinity." *Annals of Science* 18 (1962): 177–94.

Eamon, William. "New Light on Robert Boyle and the Discovery of Colour Indicators." *Ambix* 27 (1980): 204–9.

———. *Science and the Secrets of Nature: Books of Secrets in Medieval and Early Modern Culture*. Princeton: Princeton University Press, 1994.

Eddy, Matthew D. *The Language of Mineralogy: John Walker, Chemistry, and the Edinburgh Medical School, 1750–1800*. Farnham, UK: Ashgate, 2008.

Eklund, Jon. "Chemical Analysis and the Phlogiston Theory, 1738–1772: Prelude to Revolution." Ph.D. dissertation, Yale University, 1971.

———. *The Incompleat Chymist; Being an Essay on the Eighteenth-Century Chemist in His Laboratory*. Smithsonian Studies in Science and Technology, no. 33. Washington, DC: Smithsonian Institution Press, 1975.

———. "Of a Spirit in the Water: Some Early Ideas on the Aerial Dimension." *Isis* 67 (1976): 527–50.

Elena, Alberto. "Baconianism in the Seventeenth-Century Netherlands: A Preliminary Survey." *Nuncius* 6 (1991): 33–47.

Elmer, Peter. "Medicine, Religion, and the Puritan Revolution." In *The Medical Revolution of the Seventeenth Century*, edited by Roger French and Andrew Wear, 10–45. Cambridge: Cambridge University Press, 1989.

Emerson, Roger L. "The Founding of the Edinburgh Medical School." *Journal of the History of Medicine* 59 (2004): 183–218.

Ercker, Lazarus. *Fleta Minor, or, the Laws of Art and Nature, in Knowing, Judging, Fining, Refining and Inlarging the Bodies of Confined Metals*. Translated by John Suffolk. London: Thomas Dawks, 1685.

Farrington, Benjamin. *The Philosophy of Francis Bacon*. Liverpool: Liverpool University Press, 1975.

Fichman, Martin. "French Stahlism and Chemical Studies of Air, 1750–1770." *Ambix* 18 (1971): 94–122.

Fix, Andrew C. "Angels, Devils, and Evil Spirits in Seventeenth-Century Thought: Balthasar Bekker and the Collegiants." *Journal of the History of Ideas* 50 (1989): 527–47.

———. *Prophecy and Reason: The Dutch Collegiants in the Early Enlightenment*. Princeton: Princeton University Press, 1991.

Fontenelle, Bernard. "Eloge de M. Boerhaave." In *Histoire de l'Académie royale des sciences, année 1738*, 105–16. Paris, 1740.

———. "Eloge de Nicolas Lemery." In *Histoire de l'Académie royale des sciences, année 1715*, 96–108. Amsterdam, 1719.

Fors, Hjalmar. "Occult Traditions and Enlightened Science: The Swedish Board of Mines as an Intellectual Environment, 1680–1760." In *Chymists and Chymistry: Studies in the History of Alchemy and Early Modern Chemistry*, edited by Lawrence M. Principe, 239–52. Sagamore Beach, MA: Chemical Heritage Foundation and Science History Publications, 2007.

Foster, M. *Lectures on the History of Physiology during the 16th, 17th, and 18th Centuries*. New York: Dover, 1970.

Frank, Robert G. *Harvey and the Oxford Physiologists: A Study of Scientific Ideas*. Berkeley: University of California Press, 1980.

French, Roger. "Harvey in Holland: Circulation among the Calvinists." In *The Medical Revolution of the Seventeenth Century*, edited by Roger French and Andrew Wear, 46–86. Cambridge: Cambridge University Press, 1989.

Frijhoff, W. Th. M. *La Société néederlandaise et ses gradués, 1575–1818*. Amsterdam: APA/Holland University Press, 1981.

———. "Medical Education and Early Modern Dutch Medical Practitioners: Towards a Critical Approach." In *The Task of Healing: Medicine, Religion, and Gender in England and the Netherlands, 1450–1800*, edited by Hilary Marland and Margaret Pelling, 205–20. Rotterdam: Erasmus Publishing, 1996.

Galen. *Medicorum Graecorum Opera . . . Claudi Galeni*. Edited by Carolus Gottlob Kühn. Leipzig: Car. Cnoblochii, 1824.

Galison, Peter. "History, Philosophy, and the Central Metaphor." *Science in Context* 2 (1988): 197–212.

Garrison, James W. "Newton and the Relation of Mathematics to Natural Philosophy." *Journal of the History of Ideas* 48 (1987): 609–27.

Gascoigne, John. "A Reappraisal of the Role of the Universities in the Scientific Revolution." In *Reappraisals of the Scientific Revolution*, edited by David C. Lindberg and Robert S. Westman, 207–60. Cambridge: Cambridge University Press, 1990.

Gaukroger, Stephen. *Francis Bacon and the Transformation of Early-Modern Philosophy*. New York: Cambridge University Press, 2001.

Gelman, Zahkar E. "Angelo Sala, an Iatrochemist of the Late Renaissance." *Ambix* 41 (1994): 141–60.

Geoffroy, Étienne-François. "Des supercheries concernent la Pierre Philosophale." In *Mémoires de l'Académie royale des sciences, année 1722*, 372–88. Paris, 1777.

———. "Eclaircissemens sur la table insérée dans les mémoires de 1718, concernent les rapports observés entre différentes substances." In *Mémoires de l'Académie royale des sciences, année 1720*, 200–219. Paris, 1777.

———. "Experiences sur les métaux, faites avec le verre ardent au Palais Royal." In *Mémoires de l'Académie royale des sciences, année 1709*, 205–22. Paris, 1777.

———. "Manière de recomposer le souffre commun par la réunion de ses principes." In *Mémoires de l'Académie royale des sciences, année 1704*, 384–94. Amsterdam, 1747.

———. "Table des différents rapports observés en chymie entre différentes substances." In *Mémoires de l'Académie royale des sciences, année 1718*, 256–69. Paris, 1777.

Geyer-Kordesch, Johanna. "Georg Ernst Stahl's Radical Pietist Medicine and Its Influence on the German Enlightenment." In *The Medical Enlightenment of the Eighteenth Century*, edited by Andrew Cunningham and Roger French, 67–87. Cambridge: Cambridge University Press, 1990.

———. "Passions and the Ghost in the Machine: Or What Not to Ask about Science in Seventeenth- and Eighteenth-Century Germany." In *The Medical Revolution of the Seventeenth Century*, edited by Roger French and Andrew Wear, 145–63. Cambridge: Cambridge University Press, 1989.

Gibbs, F. W. "Boerhaave's Chemical Writings." *Ambix* 6 (1958): 117–35.

———. "Peter Shaw and the Revival of Chemistry." *Annals of Science* 7 (1951): 212–32.

Gibbs, G. C. "The Role of the Dutch Republic as the Intellectual Entrepôt of Europe in the Seventeenth and Eighteenth Centuries." *Bijdragen en Mededelingenden Treffende de Geschiedenis der Nederlanden* 86 (1971): 223–49.

Gilbert, Neal W. *Renaissance Concepts of Method*. New York: Columbia University Press, 1960.

Goldman, Harry S. "Weber's Ascetic Practices of the Self." In *Weber's "Protestant Ethic,"* edited by H. Lehman and G. Roth, 161–77. Cambridge: Cambridge University Press, 1993.

Golinski, Jan V. "Chemistry and the Scientific Revolution: Problems of Language and Communication." In *Reappraisals of the Scientific Revolution*, edited by David C. Lindberg and Robert S. Westman, 367–96. New York: Cambridge University Press, 1990.

———. "'Fit Instruments': Thermometers in Eighteenth-Century Chemistry." In *Instruments and Experimentation in the History of Chemistry*, edited by Frederic L. Holmes and Trevor H. Levere, 185–210. Cambridge: MIT Press, 2000.

———. "A Noble Spectacle: Phosphorous and the Public Cultures of Science in the Early Royal Society." *Isis* 80 (1989): 11–39.

———. "Peter Shaw: Chemistry and Communication in Augustan England." *Ambix* 30 (1983): 19–29.

———. *Science as Public Culture: Chemistry and Enlightenment in Britain, 1760–1820*. Cambridge: Cambridge University Press, 1992.

———. "Utility and Audience in Eighteenth-Century Chemistry: Case Studies of William Cullen and Joseph Priestly." *British Journal for the History of Science* 21 (1988): 1–31.

Grafton, Anthony. "Civic Humanism and Scientific Scholarship at Leiden." In *The University and the City: From Medieval Origins to the Present*, edited by Thomas Bender, 59–78. Oxford: Oxford University Press, 1988.

Grant, Edward. *The Foundations of Modern Science in the Middle Ages*. Cambridge: Cambridge University Press, 1996.

Green, Robert W., ed. *Protestantism, Capitalism, and Social Science: The Weber Thesis Controversy*. Lexington, MA: D. C. Heath, 1973.

Greenaway, Frank. "Boerhaave's Influence on Some 18th-Century Chemists." In *Boerhaave and His Time*, edited by G. A. Lindeboom, 102–13. Leiden: Brill, 1970.

———. "The Early Development of Analytical Chemistry." *Endeavour* 21 (1962): 91–97.

Guerlac, Henry. "The Continental Reputation of Stephen Hales." *Archives Internationales d'Histoire des Sciences* 4 (1951): 393–404.

———. "Guy de la Brosse and the French Paracelsians." In *Science, Medicine, and Society in the Re-*

naissance: Essays to Honor Walter Pagel, edited by Allen G. Debus, 1:177–99. New York: Science History Publications, 1972.

———. *Lavoisier—The Crucial Year: The Background and Origin of His First Experiments on Combustion in 1772*. 1961; reprint, New York: Gordon and Breach, 1990.

———. "Newton and the Method of Analysis." In *Dictionary of the History of Ideas*, edited by P. P. Weiner, 3:378–91. New York: Scribner, 1973.

Guerrini, Anita. "Anatomists and Entrepreneurs in Early Eighteenth-Century London." *Journal of the History of Medicine* 59 (2004): 219–39.

———. "Archibald Pitcarne and Newtonian Medicine." *Medical History* 31 (1987): 70–83.

———. "Chemistry Teaching at Oxford and Cambridge, circa. 1700." In *Alchemy and Chemistry in the Sixteenth and Seventeenth Centuries*, edited by P. Rattansi and A. Clericuzio, 183–99. Dordrecht: Kluwer, 1994.

———. "The Tory Newtonians: Gregory, Pitcarne and Their Circle." *Journal of British Studies* 25 (1986): 288–311.

———. "The Varieties of Mechanical Medicine: Borelli, Malpighi, Bellini and Pitcarne." In *Marcello Malpighi: Anatomist and Physician*, edited by Domenico Bertoloni Meli, 111–28. Florence: Olschki, 1997.

Gunnoe, Jr., Charles D. "Erastus and Paracelsianism: Theological Motifs in Thomas Erastus' Rejection of Paracelsian Natural Philosophy." In *Reading the Book of Nature: The Other Side of the Scientific Revolution*, edited by Allen G. Debus and Michael Walton, 45–66. Kirksville, MO: Sixteenth Century Press, 1998.

Guthrie, Douglas. "The Influence of the Leyden School on Scottish Medicine." *Medical History* 3 (1959): 108–22.

Hackman, W. D. "The Growth of Science in the Netherlands in the Seventeenth and Early Eighteenth Centuries." In *The Emergence of Science in Western Europe*, edited by Maurice Crosland, 89–109. London: Macmillan, 1975.

Hales, Stephen. *Vegetable Staticks, or An Account of Statical Experiments on the Sap of Vegetables*. London: W. and J. Innys and T. Woodward, 1727.

Hall, A. Rupert, and Marie Boas Hall, eds. *The Correspondence of Henry Oldenburg*. 13 vols. Madison: University of Wisconsin Press, 1965–86.

Hall, Marie Boas. "Acid and Alkali in Seventeenth-Century Chemistry." *Archives Internationales d'Histoire des Sciences* 9 (1956): 13–28.

———. *Robert Boyle and Seventeenth-Century Chemistry*. Cambridge: Cambridge University Press, 1958.

Hannaway, Owen. *The Chemists and the Word: The Didactic Origins of Chemistry*. Baltimore: Johns Hopkins University Press, 1975.

———. "Laboratory Design and the Aim of Science: Andreas Libavius versus Tycho Brahe." *Isis* 77 (1986): 585–610.

———. "Lemery, Nicolas," In *Dictionary of Scientific Biography*, 8:172–75. New York: Scribner, 1975.

Heilbron, John L. *Elements of Early Modern Physics*. Berkeley: University of California Press, 1982.

Heimann, P. M. "'Nature as Perpetual Worker': Newton's Aether in Eighteenth-Century Natural Philosophy." *Ambix* 20 (1973): 1–25.

Heninger, J. "Some Botanical Activities of Herman Boerhaave, Professor of Botany and Director of the Botanic Garden at Leiden." *Janus* 58 (1971): 1–78.

Henninger-Voss, Mary. "How the "New Science" of Cannons Shook up the Aristotelian Cosmos." *Journal of the History of Ideas* 63 (2002): 371–97.

Hermannus, Paulus. *Florae Lugduno Batavae Flores; sive, Enumeratio stirpum Horti Lugduno-Batavi methodo, naturae vestigiis insistente, dispositarum, nunc primum in lucem editorum opera Lotharis Zumbach*. Leiden, 1690.

Hippocrates. *Hippocratic Writings*. Translated by J. Chadwick and W. N. Mann. New York: Penguin, 1978.

Hollandus, Isaac. *Tractatus de Urine*. In *Theatrum Chymicum*, vol. 6., edited by Johann Jacob Heilmann. Strassburg [Argentorati]: Eberhard Zetzner, 1661.

Holmes, Frederic L. "Analysis by Fire and Solvent Extraction: The Metamorphosis of a Tradition." *Isis* 62 (1971): 129–48.

———. "Argument and Narrative in Scientific Writing." In *The Literary Structure of Scientific Argument*, edited by Peter Dear, 164–81. Philadelphia: University of Pennsylvania Press, 1991.

———. "The Communal Context for Étienne-François Geoffroy's '*Table des Rapports*.'" *Science in Context* 9 (1996): 280–311.

———. *Eighteenth-Century Chemistry as an Investigative Enterprise*. Berkeley: Office for the History of Science and Technology, University of California, 1989.

———. "Investigation and Pedagogical Style in French Chemistry at the End of the 17th Century." *Historical Studies the Physical and Biological Science* 34 (2004): 277–309.

Homberg, Wilhelm. "Essai de l'analyse du Souphre Commun." In *Mémoires de l'Académie royale des sciences, année 1703*, 39–51. Amsterdam, 1739.

——— "Essais de chimie. Article premier. Des principes de la chimie in général." In *Mémoires de l'Académie royale des sciences, année 1702*, 44–48. Amsterdam, 1737.

———. "Observation sur le fer au verre ardent." In *Mémoires de l'Académie royale des sciences, année 1706*, 199–207. Amsterdam, 1708.

———. "Observations faites par le môyen du verre ardent." In *Mémoires de l'Académie royale des sciences, année 1702*, 197–209. Amsterdam, 1737.

———. "Suite de l'article des essais de chymie." In *Mémoires de l'Académie royale des sciences, année 1706*, 336–51. Amsterdam, 1708.

Home, R. W. "Out of a Newtonian Straitjacket: Alternative Approaches to Eighteenth-Century Physical Science." *Studies in the Eighteenth Century* 4 (1979): 234–49.

Hooykaas, R. "Die Elementenlehre der Iatrochemiker." *Janus* 41 (1937): 1–28.

———. "Die Elementenlehre des Paracelsus." *Janus* 39 (1935): 175–88.

Hotson, Howard. *Commonplace Learning: Ramism and Its German Ramifications, 1543–1630*. Oxford: Oxford University Press, 1997.

Hoving-Ebels, B. "Theorie en therapie. Antonius Deusing (1612–1666) en de koorts." In *Medische gescheidenis in regionaal perspectif: Groningen 1500–1900*, edited by Frank Huisman and Catrien Santing, 49–68. Rotterdam: Erasmus, 1997.

Howard, Rio. "Guy de la Brosse and the Jardin des Plantes in Paris." In *The Analytic Spirit: Essays in the History of Science in Honor of Henry Guerlac*, edited by Harry Woolf, 195–224. Ithaca: Cornell University Press, 1981.

Hubicki, Wlodzimierz. "Alexander von Suchten." *Sudhoffs Archiv* 44, no. 1 (1960): 54–63.

Hughes, J. Trevor. *Thomas Willis, 1621–1675: His Life and Work*. London: Royal Society of Medicine Services, 1991.

Hunter, Michael. "Alchemy, Magic, and Moralism in the Thought of Robert Boyle." *British Journal for the History of Science* 23 (1990): 387–410.

Inkster, Ian. "The Public Lecture as an Instrument of Science Education for Adults: The Case of Great Britain, c. 1750–1850." *Paedogogica Historica* 20 (1980): 80–107.

Israel, Jonathan. *The Dutch Republic: Its Rise, Greatness, and Fall, 1477–1806*. Oxford: Clarendon Press, 1995.

———. *Radical Enlightenment: Philosophy and the Making of Modernity, 1650–1750*. Oxford: Oxford University Press, 2001.

———."William III and Toleration." In *From Persecution to Toleration: The Glorious Revolution in England*, edited by Ole Peter Grell, Jonathan I. Israel, and Nicholas Tyacke, 129–70. Oxford: Clarendon, 1991.

Jardine, Nicholas. "Epistemology of the Sciences." In *The Cambridge History of Renaissance Philos-*

ophy, edited by Charles B. Schmitt et al., 686–711. Cambridge: Cambridge University Press, 1988.

Jevons, F. R. "Boerhaave's Biochemistry." *Medical History* 6 (1962): 343–62.

Johns, Adrian. *The Nature of the Book: Print and Knowledge in the Making*. Chicago: University of Chicago Press, 1998.

Johnston, Ian. *Galen: On Diseases and Symptoms*. Cambridge: Cambridge University Press, 2006.

Joly, Bernard. "Alchemie et rationalité: La question des critères de démarcation entre chimie et alchemie au XVIIe siècle." *Sciences et Techniques en Perspective* 31 (1995): 93–107.

———. "L'alkahest, dissolvant universel ou quand la théorie rend pensible un pratique impossible." *Revue d'Histoire des Sciences* 49 (1996): 305–44.

———. "Quarrels between Étienne François Geoffroy and Louis Lémery at the Académie Royale des Sciences in the Early Eighteenth Century: Mechanism and Alchemy." In *Chymists and Chymistry: Studies in the History of Alchemy and Early Modern Chemistry*, edited by Lawrence M. Principe, 203–14. Sagamore Beach, MA: Chemical Heritage Foundation and Science History Publications, 2007.

Jorissen, W. P. "Gaubius, (Hieronymus) David." In *Nieuw Nederlandsch Biografisch Woordenboek*, edited by P. C. Molhuysen and P. J. Blok, vol. 3. Leiden: Sijthoff's Uitgevers-Mattschappij, 1914.

Jurriaanse, M. W. *The Founding of Leyden University*. Translated by J. Brotherhood. Leiden: Brill, 1965.

Kaau, Hermannus. *Dissertatio Inauguralis de Argent Vivo*. Leiden: Apud Isaacum Severnium, 1729.

Kaiser, David, ed. *Pedagogy and the Practice of Science: Historical and Contemporary Perspectives*. Cambridge: MIT Press, 2005.

Kegel-Brinksgreve, E., and A. M. Luyendijk-Elshout, eds. *Boerhaave's Orations: Translated, with Introductions and Notes*. Leiden: Brill, 1983.

Kent, A., and O. Hannaway. "Some New Considerations of Beguin and Libavius." *Annals of Science* 16 (1960): 241–50.

Kerker, Milton. "Herman Boerhaave and the Development of Pneumatic Chemistry." *Isis* 46 (1955): 36–49.

Kim, Mi Gyung. *Affinity, That Elusive Dream: A Genealogy of the Chemical Revolution*. Cambridge: MIT Press, 2003.

———. "The 'Instrumental' Reality of Phlogiston." *HYLE: International Journal for the Philosophy of Chemistry* 14 (2008): 27–51.

King, Lester. *The Medical World of the Eighteenth Century*. Chicago: Chicago University Press, 1958.

———. "Medicine in 1695: Friederich Hoffman's *Fundamenta Medicinae*." *Bulletin for the History of Medicine* 43 (1969): 17–29.

———. *The Road to Medical Enlightenment, 1650–1695*. New York: American Elsevier, 1970.

———. "Stahl and Hoffman: A Study in Eighteenth Century Animism." *Journal of the History of Medicine* 19 (1964): 118–30.

Klein, Ursula. "Apothecary-Chemists in Eighteenth-Century Germany." In *New Narratives in Eighteenth-Century Chemistry. Archimedes*, edited by Lawrence M. Principe, 18:97–137. Dordrecht: Springer, 2007.

———. "Blending Technical Innovation and Learned Knowledge: The Making of Ethers." In *Materials and Expertise in Early Modern Europe: Between Market and Laboratory*, edited by Ursula Klein and E. C. Spary, 125–57. Chicago: University of Chicago Press, 2010.

———. "E. F. Geoffroy's Table of Different 'Rapports' Observed between Different Chemical Substances—A Reinterpretation." *Ambix* 42 (1995): 79–100.

———. "Experimental History and Herman Boerhaave's Chemistry of Plants." *Studies in History and Philosophy of Science, Part C: Studies in History and Philosophy of Biological and Biomedical Sciences* 34 (2003): 533–67.

———. "The Laboratory Challenge: Some Revisions of the Standard View of Early Modern Experimentation." *Isis* 99 (2008): 769–82.

———. "Nature and Art in Seventeenth-Century French Chemical Textbooks." In *Reading the Book of Nature: The Other Side of the Scientific Revolution*, edited by Allen G. Debus and Michael Walton, 239–50. Kirksville, MO: Sixteenth Century Press, 1998.

Klein, Ursula, and Wolfgang Lefèvre. *Materials in Eighteenth-Century Science: A Historical Ontology*. Cambridge: MIT Press, 2007.

Klever, W. N. A. "Burchard de Volder (1643–1709), a Crypto-Spinozist on a Leiden Cathedra." *Lias* 15 (1988): 191–241.

Knight, David. *Ideas in Chemistry: A History of Science*. New Brunswick, NJ: Rutgers University Press, 1992.

Knight, Harriet, and Michael Hunter. "Robert Boyle's *Memoirs for the Natural History of Human Blood* (1684): Print, Manuscript and Impact of Baconianism on Seventeenth-Century Medical Science." *Medical History* 51 (2007): 145–64.

Knoeff, Rina. "Chemistry, Mechanics, and the Making of Anatomical Knowledge: Boerhaave vs. Ruysch on the Nature of the Glands." *Ambix* 53 (2006): 201–19.

———. *Herman Boerhaave (1668–1738): Calvinist Chemist and Physician*. Amsterdam: Koninklijke Nederlandse Akademie van Wetenschappen, 2002.

———. "The Making of a Calvinist Chemist: Herman Boerhaave, God, Fire and Truth." *Ambix* 48 (2001): 102–11.

———. "Practicing Chemistry "After the Hippocratical Manner": Hippocrates and the Importance of Chemistry in Boerhaave's Medicine." In *New Narratives in Eighteenth-Century Chemistry. Archimedes*, edited by Lawrence Principe, 18:63–76. Dordrecht: Springer, 2007.

Kristeller, Paul Oskar. "The Modern System of the Arts." In *Renaissance Thought and the Arts: Collected Essays*, 163–227. Princeton: Princeton University Press, 1990.

Kroon, J. E. "Boerhaave as Professor-Promoter." *Janus* 21 (1918): 291–311.

Kuhn, Thomas. "Robert Boyle and Structural Chemistry in the Seventeenth Century." *Isis* 43 (1952): 12–36.

———. *The Structure of Scientific Revolutions*. 2nd ed. Chicago: University of Chicago Press, 1970.

———. "The Essential Tension: Tradition and Innovation in Scientific Research?" In *The Essential Tension: Selected Studies in Scientific Tradition and Change*, 225–39. Chicago: University of Chicago Press, 1977.

Lavoisier, Antoine-Laurent. *Elements of Chemistry, in a New Systematic Order, containing All the Modern Discoveries*. Translated by Robert Kerr. New York: Dover, 1965.

———. *Opuscules physiques et chimiques*. Paris: Deterville, 1801.

Lawrence, Susan C. "Educating the Senses: Students, Teachers and Medical Rhetoric in Eighteenth-Century London" In *Medicine and the Five Senses*, edited by W. F. Bynum and Roy Porter, 154–78. New York: Cambridge University Press, 1993.

LeClerc, Jean. "Eloge de feu Mr. De Volder, professor en philosophie et aux mathematiques, dans l'Académie de Leide." *Bibliothèque Choise* (Amsterdam) 18 (1709): 346–401.

Lehman, Christine. "Between Commerce and Philanthropy: Chemistry Courses in Eighteenth-Century Paris." In *Science and Spectacle in the European Enlightenment*, edited by Bernadette Bensaude-Vincent and Christine Blondel, 103–16. Aldershot, UK: Ashgate, 2008.

———. "Mid-Eighteenth-Century Chemistry as Seen through Student Notes from the Courses of Gabriel-François Venel and Guillaume-François Rouelle." *Ambix* 56 (2009): 163–89.

Lehman, Harmun, and Guenther Roth, eds. *Weber's "Protestant Ethic": Origins, Evidence, Contexts*. Cambridge: Cambridge University Press, 1993.

Lemery, Nicolas. *Cours de chymie, contenant la manière de faire les operations qui sont en usage dans la medicine, par une methode facile*. 10th ed. Paris: Chez Jean-Baptiste Delespina, 1713.

———. *Traité de l'antimoine*. Paris, 1707.

Le Mort, Jacobus. *Chymia Verae Nobilitas et Utilitas, in Physica Corpusculari, Theoria Medica, Ejusque Materia et Signis ad Majorem Perfectionem Deducendis*. Leiden: Fredericum Haaring and Cornelium Boutesteyn, 1696.

———. *Compendium Chymicum, Demonstrans Experimentis & Rationibus brevem & facilem methodum operationes accurate & succincte ad finem perducendi.* Leiden: Apud Felicem Lopez, 1682.

———. *Ignorantia circa Chemiam et Universam Scientiam Naturalem Detecta ab Marggravio.* Leiden, 1687.

———. *Pharmacia Medico-Physico, Rationibus et Experimentis Instructa, Accuratione Methodo Adornata.* Leiden: Van der Aa, 1684.

Lesky, Erna. "Albrecht von Haller, Gerard van Swieten, und Boerhaaves Erbe." *Gesnerus* 15 (1958): 120–40.

Lesky, Erna, and Adam Wandruszka, eds. *Gerard van Swieten und seine Zeit.* Vienna: Böhlaus, 1973.

Levere, Trevor H. "Relations and Rivalry: Interactions between Britain and the Netherlands in Eighteenth-Century Science and Technology." *History of Science* 9 (1970): 42–53.

Libavius, Andreas. *Alchemia.* Frankfurt: Excudebat Johannes Saurius, impensis Pertri Kopffij, 1597.

Lindeboom, G. A. *Bibliographia Boerhaaviana.* Leiden: Brill, 1959.

———, ed. *Boerhaave and His Time.* Leiden: Brill, 1970.

———. "Boerhaave's Concept of the Basic Structure of the Body." *Clio Medica* 5 (1970): 203–8.

———, ed. *Boerhaave's Correspondence.* 3 vols. Leiden: Brill, 1962–70.

———. "Boerhaave's Impact on the Relation between Chemistry and Medicine." *Clio Medica* 7 (1972): 271–78.

———. "David en Nicholaas Stam, apothekers te Leiden." *Pharmaceutisch Weekblad* 108 (1973): 153–60.

———. *Dutch Medical Biography.* Amsterdam: Rodopi, 1984.

———. "Frog and Dog: Physiological Experiments at Leiden during the Seventeenth Century." In *Leiden University in the Seventeenth Century*, edited by Lunsingh Scheurleer and Posthumus Meyjes, 279–93. Leiden: Universitaire Pers Leiden/Brill, 1975.

———. *Herman Boerhaave: The Man and His Work.* London: Methuen, 1968.

———. "Linnaeus and Boerhaave." *Janus* 46 (1957): 264–74.

———. "Medical Education in the Netherlands, 1575–1750." In *The History of Medical Education*, edited by C. D. O'Malley, 201–16. Berkeley: University of California Press, 1970.

———. "Pitcarne's Leyden Interlude Described from the Documents." *Annals of Science* 19 (1963): 273–84.

Lindemann, Mary. *Medicine and Society in Early Modern Europe.* Cambridge: Cambridge University Press, 1999.

Lloyd, G. E. R. "The Hippocratic Question." In *Methods and Problems in Greek Science: Selected Papers*, 194–223. Cambridge: Cambridge University Press, 1991.

Lohr, Charles H. "Latin Aristotelianism and the Seventeenth-Century Calvinist Theory of Scientific Method." In *Method and Order in Renaissance Philosophy of Nature*, edited by Daniel A. DiLiscia, Eckhard Kessler, and Charlotte Meuthen, 369–80. Aldershot, UK: Ashgate, 1997.

Long, Pamela O. *Openness, Secrecy, Authorship: Technical Arts and the Culture of Knowledge from Antiquity to the Renaissance.* Baltimore: Johns Hopkins University Press, 2001.

Lonie, Iain M. "Hippocrates as Iatromechanist." *Medical History* 25 (1981): 113–50.

———. "The 'Paris Hippocrates': Teaching and Research in Paris in the Second Half of the Sixteenth Century." In *The Medical Renaissance of the Sixteenth Century*, edited by Andrew Wear, Roger K. French, and Ian M. Lonie, 155–74. Cambridge: Cambridge University Press, 1985.

Love, Rosaleen. "Herman Boerhaave and the Element-Instrument Concept of Fire." *Annals of Science* 31 (1974): 547–59.

———. "Some Sources of Herman Boerhaave's Concept of Fire." *Ambix* 19 (1972): 157–74.

Lundgren, Anders, and Bernadette Bensaude-Vincent, eds. *Communicating Chemistry: Textbooks and Their Audiences, 1789–1939.* Canton, MA: Science History Publications.

Lunsingh Sheurleer, Th. H. "Un amphithéâtre d'anatomie moralisée." In *Leiden University in the Seventeenth Century*, edited by Lunsingh Scheurleer and Posthumus Meyjes, 217–77. Leiden: Universitaire Pers Leiden/Brill, 1975.

Lunsingh Scheurleer, Th. H., and G. H. M. Posthumus Meyjes, eds. *Leiden University in the Seventeenth Century: An Exchange of Learning*. Leiden: Universitaire Pers Leiden/Brill, 1975.

Luyendijk-Elshout, Antonie M. "Mechanisme contra vitalisme de school van Herman Boerhaave en de Beginselen van het Leven." *Tijdschrift voor de Geschiedenis der Geneeskunde, Natuurwetenschappen, Wiskunde en Techniek* 5 (1982): 16–26.

———. "*Oeconomia Anamalis*, Pores, and Particles: The Rise and Fall of the Medical Philosophical School of Theodoor Craanen (1621–1690)." In *Leiden University in the Seventeenth Century*, edited by Lunsingh Scheurleer and Posthumus Meyjes, 295–307. Leiden: Universitaire Pers Leiden/Brill, 1975.

Macquer, Pierre-Joseph. *Dictionnaire de chymie*. Neuchatel: Société Typographique, 1789.

Marggraf, Christian. *Jacobi le Mort, Pseudochemici et Ratiocinatoris Dupoindiarii Ignorantia circa Chemiam et Universam Scientarum Naturalem*. Leiden, 1687.

Mauskopf, Seymour. "Reflections: 'A Likely Story.'" In *New Narratives in Eighteenth-Century Chemistry*, edited by Lawrence M. Principe, 177–93. Dordrecht: Springer, 2007.

McGahagan, Thomas. "Cartesianism in the Netherlands, 1639–1676: The New Science and the Calvinist Counter-Reformation." Ph.D. dissertation, University of Pennsylvania, 1976.

McNeill, John T. *The History and Character of Calvinism*. Oxford: Oxford University Press, 1954.

Meinel, Christoph. "*Artibus Academicus Inserenda*: Chemistry's Place in Eighteenth and Nineteenth Century Universities." *History of Universities* 7 (1988): 89–115.

———. "Theory or Practice? The Eighteenth-Century Debate on the Scientific Status of Chemistry." *Ambix* 30 (1983): 121–32.

Melhado, Evan M. "Oxygen, Phlogiston, and Caloric: The Case of Guyton." *Historical Studies in the Physical Sciences* 13 (1982): 311–34.

Metzger, Hélène. *Les doctrines chimiques en France au debut du XIIe à la fin du XIIIe siècle*. Paris: Blanchard, 1923.

———. *Newton, Stahl, Boerhaave et les doctrines chimiques*. Paris: Blanchard, 1930.

Middleton, W. E. Knowles. *A History of the Thermometer and Its Use in Meteorology*. Baltimore: Johns Hopkins University Press, 1966.

Milburn, John R. "The London Evening Courses of Benjamin Martin and James Ferguson, Eighteenth-Century Lecturers on Experimental Philosophy." *Annals of Science* 40 (1983): 437–55.

Molhuysen, P. C., ed. *Bronnen tot de Geschiedenis der Leidsche Universiteit*. 7 vols. Rijks Geschiedkundige Publicatien, nos. 20, 29, 38, 45, 48, 53, and 56. 's-Gravenhage: Martinus Nijhoff, 1913–23.

Molhuysen, P. C., and P. J. Blok, eds. *Niew Nederlandsch Biografisch Woordenboek*. Leiden: A. W. Sijthoff's Uitgevers-Mattschappij, 1911.

Moran, Bruce T. *Andreas Libavius and the Transformation of Alchemy: Separating Chemical Cultures with Polemical Fire*. Sagamore Beach, MA: Science History Publications, 2007.

———. "Axioms, Essences and Mostly Clean Hands: Preparing to Teach Chemistry with Libavius and Aristotle," *Science and Education*. 15 (2006): 173–87.

———. *Chemical Pharmacy Enters the University: Johannes Hartmann and the Didactic Care of* Chymiatria *in the Early Seventeenth Century*. Madison, WI: American Institute of the History of Pharmacy, 1991.

———. *Distilling Knowledge: Alchemy, Chemistry and the Scientific Revolution* Cambridge: Harvard University Press, 2005.

Morley, Christopher Love. *Collectanea Chymica Leydensia, id est, Maetsiana, Margraviana, et le Mortiana; Silicet trium in Academia Lugduno-Batava facultatis chimiae*. Leiden: Apud Henricum Drummond, sumptibus J. A. de la Font, 1684.

Multhauf, Robert P. *The Origins of Chemistry*. New York: Franklin Watts, 1966.

———. "The Significance of Distillation in Renaissance Medical Chemistry." *Bulletin of the History of Medicine* 30 (1956): 329–46.

Musaeum Hermeticum Reformatum et Amplificatum, omnes sopho-spagyricae artis discipulos fidelis-

sime erudiens, quo pacto Summa illa veraque Lapidis Philosophici Medicina, qua res omnes qualemcunque defectum patientes, iinstaurantur, inveniri & haberi queat. Frankfurt: Apud Hermannum a Sande, 1678.

Newman, William R. "Alchemical Symbolism and Concealment: The Chemical House of Libavius." In *The Architecture of Science*, edited by Peter Galison and Emily Thompson, 61–77. Cambridge: MIT Press, 1998.

———. *Atoms and Alchemy: Chymistry and the Experimental Origins of the Scientific Revolution.* Chicago: University of Chicago Press, 2006.

———. "The Authorship of the *Introitus Apertus ad Occulsum Regis Palatium.*" In *Alchemy Revisited: Proceedings of the International Conference on the History of Alchemy at the University of Groningen, 17–18 April 1989*, edited by Z. R. M. W. von Martels, 139–44. Leiden: Brill, 1990.

———. *Gehennical Fire: The Lives of George Starkey, an American Alchemist in the Scientific Revolution.* Cambridge: Harvard University Press, 1994.

———. "Newton's Theory of Metallic Generation in the Previously Neglected Text 'Humores minerales continuo decidunt.'" In *Chymists and Chymistry: Studies in the History of Alchemy and Early Modern Chemistry*, edited by Lawrence M. Principe, 89–99. Sagamore Beach, MA: History Science Publications, 2007.

———. "Prophecy and Alchemy: The Origins of Eirenaeus Philalethes." *Ambix* 37 (1990): 102–6.

———. *The Summa Perfectionis of Pseudo-Geber*. Leiden: Brill, 1991.

———. "Technology and Alchemical Debate in the Late Middle Ages." *Isis* 80 (1989): 423–45.

Newman, William R., and Lawrence M. Principe. *Alchemy Tried in the Fire: Starkey, Boyle and the Fate of Helmontian Chymistry*. Chicago: University of Chicago Press, 2002.

———. "Alchemy vs. Chemistry: The Etymological Origins of a Historiographic Mistake." *Early Science and Medicine* 3 (1998): 32–65.

Newton, Isaac. *Opticks, or a Treatise on the Reflections, Refractions, Influctions and Colours of Light.* New York: Cosimo, 2007.

Norris, John A. "Early Theories of Aqueous Mineral Genesis in the Sixteenth Century." *Ambix* 54 (2007): 69–86.

———. "The Mineral Exhalation Theory of Metallogenesis in Pre-Modern Mineral Science." *Ambix* 53 (2006): 43–65.

Novak, Alfred. "Boerhaave: Three Chairs to Oblivion." *Bioscience*. 21 (1971): 479–82.

Nuck, Anton. *De Ductu Salivali Novo, Saliva, Ductibus Oculorum Aquosis, et Humore Oculi Aqueo.* Leiden: Petrum vander Aa, 1685.

Nutton, Vivian. *Ancient Medicine*. New York: Routledge, 2005.

———. "Hippocrates in the Renaissance." In *Die Hippokratischen Epidemien: Theorie—Praxis—Tradition*, edited by Gerhard Baader and Rolf Winau, 420–39. Stuttgart: Franz Steiner Verlag, 1990.

Olesko, Kathryn M. *Physics as a Calling: Discipline and Practice in the Köningsberg Seminar for Physics*. Ithaca: Cornell University Press, 1991.

Oldroyd, David. "An Examination of G. E. Stahl's *Philosophical Principles of Universal Chemistry.*" *Ambix* 20 (1973): 36–52.

———. "Some Phlogistic Mineralogical Schemes, Illustrative of the Evolution of the Concept of 'Earth' in the 17th and 18th Centuries." *Annals of Science* 31 (1974): 269–305.

Ong, Walter J. *Ramus, Method, and the Decay of Dialogue*. Cambridge: Harvard University Press, 1958.

Otterspeer, Willem. *Het bolwerk van de vrijheid: De Leidse universiteit, 1575–1672*. Amsterdam: Bert Bakker, 2000.

Pagel, Walter. *Joan Baptista van Helmont: Reformer of Science and Medicine*. Cambridge: Cambridge University Press, 1982.

———. *Paracelsus: An Introduction to Philosophical Medicine in the Era of the Renaissance*. New York: S. Karger, 1958.

Paracelsus. *Four Treatises*. Edited by Henry E. Sigerist. Baltimore: Johns Hopkins University Press, 1941.

———. *Sämtliche Werke*. 14 vols. Edited by Karl Sudhoff and Wilhelm Matthiessen. Munich: R. Oldenbourg, 1922–34.

Partington, John R. *A History of Chemistry*. 4 vols. New York: St. Martin's Press, 1961–70.

Patterson, T. S. "Jean Beguin and His *Tyrocinium Chymicum*." *Annals of Science* 2 (1937): 243–98.

Penning, C. P. J. "De Promotie van Boerhaave te Harderwijk." *Nederlandsch Tijdschrift voor Geneeskunde* 82 (1938): 4895–99.

Pereira, Michela. *The Alchemical Corpus Attributed to Raymond Lull*. Warburg Institute Surveys and Texts, 18. London: University of London Press, 1989.

Pérez-Ramos, Antonio. *Francis Bacon's Idea of Science and the Maker's Knowledge Tradition*. Oxford: Clarendon Press, 1988.

Philalethes, Eirenaeus. *Introitus Apertus ad Occlusum Regis Palatium, Auttore Anonymo Philaletha Philosopho*. In *Musaeum Hermeticum Reformatum et Amplificatum*, 647–99. Frankfurt: Hermannum a Sande, 1678.

Porta, Paulo A. "'*Summus atque felicissimus salium*': The Medical Relevance of the Liquor Alkahest." *Bulletin of the History of Medicine* 76 (2002): 1–29.

Porter, Roy. "Science, Provincial Culture, and Public Opinion in Enlightenment England." *British Journal of Eighteenth-Century Studies* 3 (1980): 20–46.

Posthumus Meyjes, G. H. M. "Le Collège Wallon." In *Leiden University in the Seventeenth Century*, edited by Lunsingh Scheurleer and Posthumus Meyjes, 111–35. Leiden: Universitaire Pers Leiden/Brill, 1975.

Powers, John C. "'*Ars sine arte*': Nicholas Lemery and the End of Alchemy in Eighteenth-Century France." *Ambix* 45 (1998): 163–89.

———. "Chemistry Enters the University: Herman Boerhaave and the Reform of the Chemical Arts." *History of Universities* 25 (2006): 77–116.

———. "Chemistry without Principles: Herman Boerhaave on Instruments and Elements." In *New Narratives in Eighteenth-Century Chemistry*, edited by Lawrence M. Principe, 45–61. Dordrecht: Springer, 2007.

———. "Scrutinizing the Alchemists: Herman Boerhaave and the Testing of Chymistry." In *Chymists and Chymistry: Studies in the History of Alchemy and Early Modern Chemistry*, edited by Lawrence M. Principe, 227–38. Sagamore Beach, MA: Chemical Heritage Foundation and Science History Publications, 2007.

Principe, Lawrence M. *The Aspiring Adept: Robert Boyle and His Alchemical Quest*. Princeton: Princeton University Press, 1998.

———. "Diversity in Alchemy: The Case of Gaston 'Claveus' DuClo, a Scholastic Mercurialist Chrysopoeian." In *Reading the Book of Nature: The Other Side of the Scientific Revolution*, edited by Allen G. Debus and Michael T. Walton, 181–200. Kirksville, MO: Sixteenth Century Press, 1998.

———, ed. *New Narratives in Eighteenth-Century Chemistry*. Dordrecht: Springer, 2007.

———. "A Revolution Nobody Noticed? Changes in Early Eighteenth-Century Chemistry." In *New Narratives in Eighteenth-Century Chemistry*, edited by Lawrence M. Principe, 1–22. Dordrecht: Springer, 2007.

———. "Robert Boyle's Alchemical Secrecy: Codes, Ciphers, and Concealments." *Ambix* 39 (1992): 63–74.

———. "Wilhelm Homberg: Chymical Corpuscularianism and Chrysopoeia in the Early Eighteenth Century." In *Late Medieval and Early Modern Corpuscular Matter Theories*, edited by Christoph Lüthy, John E. Murdoch, and William R. Newman, 535–56. Leiden: Brill, 2001.

Principe, Lawrence M., and William R. Newman. "Some Problems in the Historiography of Alchemy." In *Secrets of Nature: Astrology and Alchemy in Early Modern Europe*, edited by William R. Newman and Anthony Grafton, 385–431. Cambridge: MIT Press, 2001.

Rademaker, C. S. M. "The Famous Library of Gerardus Joannes Vossius (1577–1649)." *Lias* 23 (1996): 27–47.

Ragland, Evan R. "Experimenting with Chymical Bodies: Reinier de Graaf's Investigations of the Pancreas." *Early Science and Medicine* 13 (2008): 615–64

Rappaport, Rhoda. "G.-F. Rouelle, an Eighteenth-Century Chemist and Teacher." *Chymia* 6 (1960): 68–101.

———. "Rouelle and Stahl: The Phlogistic Revolution in France." *Chymia* 7 (1961): 73–102.

Read, John. *Humour and Humanism in Chemistry*. London: G. Bell, 1947.

———. *From Alchemy to Chemistry*. New York: Dover, 1995.

Reif, Patricia. "The Textbook Tradition in Natural Philosophy, 1600–1650." *Journal of the History of Ideas* 30 (1969): 17–32.

Risse, Guenter. "Clinical Instruction in Hospitals: The Boerhaavian Tradition in Leyden, Edinburgh, Vienna, and Pavia." *Clio Medica* 21 (1987–88): 1–19.

Roberts, Lissa. "Chemistry on Stage: G. F. Rouelle and the Theatricality of Eighteenth-Century Chemistry." In *Science and Spectacle in the European Enlightenment*, edited by Bernadette Bensaude-Vincent and Christine Blondel, 129–39. Aldershot, UK: Ashgate, 2008.

———. "The Death of the Sensuous Chemist: The 'New' Chemistry and the Transformation of Sensuous Technology." *Studies in History and Philosophy of Science* 26 (1995): 503–29.

———. "Going Dutch: Situating Science in the Dutch Enlightenment." In *The Sciences in Enlightened Europe*, edited by William Clark, Jan Golinski, and Simon Schaffer, 249–88. Chicago: University of Chicago Press, 1999.

———. "Setting the Table: The Disciplinary Development of Eighteenth-Century Chemistry as Read through the Changing Structure of Its Tables." In *The Literary Structure of Scientific Argument*, edited by Peter Dear, 99–132. Philadelphia: University of Pennsylvania Press, 1991.

Roberts, Lissa, Simon Schaffer, and Peter Dear, eds. *The Mindful Hand: Inquiry and Invention from the Late Renaissance to Early Industrialization*. Amsterdam: Koninkijke Nederlandse Akademie van Wetenschappen, 2007.

Rolfincius, Gvernerus. *Chemia in Artis Formam Redacta, Sex Libris Comprehansa*. Jena: Samuel Krebs, 1661.

Roos, Anna Marie. *The Salt of the Earth: Natural Philosophy, Medicine, and Chymistry in England, 1650–1750*. Boston: Brill, 2007.

Rothschuch, K. "Bohn, Johannes." In *Dictionary of Scientific Biography*, edited by Charles Gillispie, 2:237–38. New York: Scribner, 1973.

Rowbottom, Margaret. "The Teaching of Experimental Philosophy in England." In *Actes du XIe Congrès International d'Histoire des Sciences, Varsovie-Cracovie, 1965*, 4:46–53. Warsaw: Ossolineum, 1968.

Rowen, Herbert H. *The Princes of Orange: The Stadholders in the Dutch Republic*. New York: Cambridge University Press, 1988.

Ruestow, Edward G. *The Microscope in the Dutch Republic: The Shaping of Discovery*. Cambridge: Cambridge University Press, 1996.

———. *Physics at Seventeenth- and Eighteenth-Century Leiden: Philosophy and the New Science at the University*. The Hague: Martinus Nijhoff, 1973.

———. "The Rise of the Doctrine of Vascular Secretion in the Netherlands." *Journal of the History of Medicine* 35 (1980): 265–87.

Rupp, Jan C. C. "Matters of Life and Death: The Social and Cultural Conditions of the Rise of Anatomical Theatres, with Special Reference to Seventeenth-Century Holland." *History of Science* 28 (1990): 263–87.

Sailor, Danton B. "Moses and Atomism." *Journal of the History of Ideas* 25 (1964): 3–16.

Sargent, Rose-Mary. "Scientific Experiment and Legal Expertise: The Way of Experience in Seventeenth-Century England." *Studies in History and Philosophy of Science* 20 (1989): 19–45.

Sassen, F. L. R. "The Intellectual Climate in Boerhaave's Time." In *Boerhaave and His Time*, edited by G. A. Lindeboom, 1-16. Leiden: Brill, 1970.

Schaffer, Simon. "The Consuming Flame: Electrical Showmen and Tory Mystics in the World of Goods." In *Consumption and the World of Goods*, edited by John Brewer and Roy Porter, 484-526. London: Routledge, 1993.

———. "The Glorious Revolution and Medicine in Britain and the Netherlands." *Notes and Records of the Royal Society of London* 43 (1989): 167-88.

———. "Natural Philosophy." In *The Ferment of Knowledge: Studies in the Historiography of Eighteenth-Century Science*, edited by G. S. Rousseau and Roy Porter, 55-92. Cambridge: Cambridge University Press, 1980.

———. "Natural Philosophy and Public Spectacle in the Eighteenth Century." *History of Science* 21 (1983): 1-43.

Schama, Simon. *The Embarrassment of Riches: An Interpretation of Dutch Culture in the Golden Age*. Berkeley: University of California Press, 1988.

Schofield, Robert E. *Mechanism and Materialism: British Natural Philosophy in an Age of Reason*. Princeton: Princeton University Press, 1970.

Schulte, B. P. M. *Hermanni Boerhaave Praelectiones de Morbis Nervorum, 1730-1735: Een Medisch-Historische Studie van Boerhaave's Manuscript over Zenuwziekten*. Analecta Boerhaaviana II. Leiden: Brill, 1959.

Schultens, Albert. *Oratio Academica in Memoriam Hermanni Boerhaavi*. London, 1739.

Scopa, James P. "Boerhaave on Alchemy." *Synthesis* 4 (1979): 23-37.

Secord, James A. "Newton in the Nursery: Tom Telescope and the Philosophy of Tops and Balls, 1761-1838." *History of Science* 2 (1985): 127-51.

Senguerdius, Wolfredus. *Philosophia Naturalis, quatuor partibus primarius corporum species, affectiones, differentias, productiones, mutationes, & interitus exhibens*.2nd ed. Leiden: Daniel a Gaesbeeck, 1685.

Sennert, Daniel. *De Chymicorum cum Aristotelicis et Galenicis Consensu ac Dissensu Liber*. Wittenberg, 1629.

Shackelford, Jole. "Paracelsian and Hippocratic Theory in Petrus Severinus' Medical Philosophy." In *Reinventing Hippocrates*, edited by David Cantor, 59-88. Aldershot, UK: Ashgate, 2002.

———. "Paracelsian Uroscopy and German Chemiatric Medicine in the *Medicina Pensylvania* of George de Benneville." In *Medical Theory and Therapeutic Practice: A Transatlantic Perspective*, edited by Jürgen Helm and Renate Wilson, 13-36. Stuttgart: Franz Steiner Verlag, 2008.

———. "Tycho Brahe, Laboratory Design, and the Aim of Science: Reading Plans in Context." *Isis* 84 (1993): 211-30.

Shapin, Steven. "Pump and Circumstance: Robert Boyle's Literary Technology." *Social Studies of Science* 14 (1983): 481-520.

———. "The House of Experiment in Seventeenth-Century England." *Isis* 79 (1988): 373-404.

Shapin, Steven, and Simon Schaffer. *Leviathan and the Air-Pump: Hobbes, Boyle, and the Experimental Life*. Princeton: Princeton University Press, 1985.

Shapiro, Barbara J. "The Universities and Science in Seventeenth-Century England." *Journal of British Studies* 10 (1971): 47-82.

Shaw, Peter. *Chemical Lectures, Publickly Read at London in the Years 1731 and 1732, and Since at Scarborough, in 1733*. London: J. Shuckburgh and Tho. Osborne, 1734.

Siegfried, Robert. *From Elements to Atoms: A History of Chemical Composition*. Philadelphia: American Philosophical Society, 2002.

Sigerist, Henry E. "Boerhaave's Influence on Early American Medicine." *Nederlandsch Tijdschrift voor Geneeskunde* 82 (1938): 4822-28.

Siraisi, Nancy G. *Medieval and Early Renaissance Medicine: An Introduction to Knowledge and Practice*. Chicago: University of Chicago Press, 1990.

Smeaton, William. "Some Large Burning Lenses and Their Uses by Eighteenth-Century French and British Chemists." *Annals of Science* 4 (1987): 265–76.

Smith, Pamela H. *The Body of the Artisan: Art and Experience in the Scientific Revolution.* Chicago: University of Chicago Press, 2004.

———. *The Business of Alchemy: Science and Culture in the Holy Roman Empire.* Princeton: Princeton University Press, 1994.

———. "Science and Taste: Painting, Passions, and the New Philosophy in Seventeenth-Century Leiden." *Isis* 90 (1999): 421–61.

———. "A Sixteenth-Century Goldsmith's Workshop." In *The Mindful Hand: Inquiry and Invention from the Late Renaissance to Early Industrialization*, edited by Lissa Roberts, Simon Schaffer, and Peter Dear, 32–57. Amsterdam: Koninkijke Nederlandse Akademie van Wetenschappen, 2007.

Snelders, Henricus A. M. "Georg Ernst Stahls Phlogiston und Herman Boerhaaves Pabulum Ignis: Eine vergleichende Analyse." In *Georg Ernst Stahl (1659–1734)*, edited by Wolfram Kaiser and Arina Völker, 177–87. Hallesches Symposium, 1984. Halles: Universität Halle-Wittenberg, 1985.

Snellen, Paulus. *Disputatio Metallurgico-Physico-Medica Inauguralis de Historia Metallorum.* Leiden: Abraham Elzevier, 1707.

Stahl, Georg Ernst. *Bedenken von der Gold-Macherey.* In *Chymischer Glücks-Hafen, Oder Grosse Chymische Concordantz und Collection von Funffzehen hundert Chymischen Processen*, by J. J. Becher. Edited by Georg Ernst Stahl. Leipzig: J. P. Kraus, 1755.

———. *Fundamenta Chymiae Dogmaticae et Experimentalis.* Nuremburg: Wolfgang Manturius, 1723.

———. *Opusculum Chymico-Physico-Medicum, seu Schediasmatum a pluribus annis variis occasionibus in publicum emissorum nunc quadantenusetiam auctorum et deficientibus passim exemplaribus in unum volumen jam collectorum.* Halle, 1715.

———. *Philosophical Principles of Universal Chemistry; or, The Foundation of a Scientifical Manner of Inquiring into and Preparing the Natural and Artificial Bodies for the Uses of Life.* Translated by Peter Shaw. London: John Osborn and Thomas Longman, 1730.

———. *Zymotechnia Fundamentalis sen Fermentationis Theoria Generalis.* Halle, 1697.

Stewart, Larry. *The Rise of Public Science: Rhetoric, Technology, and Natural Philosophy in Newtonian Britain, 1660–1750.* Cambridge: Cambridge University Press, 1992.

Stolberg, Michael. "The Decline of Uroscopy in Early Modern Learned Medicine (1500–1650)." *Early Science and Medicine* 12 (2007): 313–36.

Strube, Irene. "On the Importance of the Phlogiston-Theory of G. E. Stahl (1659–1734)." In *Actes du XIe Congrès International d'Histoire des Sciences, 1965*, 4:82–84. Warsaw: Ossolineum, 1968.

Sumner, James. "Michael Combrune, Peter Shaw and Commercial Chemistry: The Boerhaavian Chemical Origins of Brewing Thermometry." *Ambix* 54 (2007): 5–29.

Sutton, Geoffrey V. *Science for a Polite Society: Gender, Culture, and the Demonstration of Enlightenment.* Boulder, CO: Westview Press, 1995.

Sylla, Edith. "Science for Undergraduates in Medical Universities." In *Science and Technology in Medieval Society*, edited by Pamela O. Long, 171–86. New York: New York Academy of Sciences, 1985.

Sylvius, Franciscus. *Opera Medica.* Amsterdam: Daniel Elsevier and Abraham Woldgang, 1679.

Taylor, F. Sherwood. *The Alchemists: Founders of Modern Chemistry.* New York: Henry Schuman, 1949.

———. "The Origin of the Thermometer." *Annals of Science* 5 (1942): 129–56.

Taylor, Georgette. "Making Out a Common Disciplinary Ground: The Role of Chemical Pedagogy in Establishing the Doctrine of Affinity at the Heart of British Chemistry." *Annals of Science* 65 (2008): 465–86.

———. "Unification Achieved: William Cullen's Theory of Heat and Phlogiston as an Example of His Philosophical Chemistry." *British Journal for the History of Science* 39 (2006): 477–501.

Thackray, Arnold. *Atoms and Powers: An Essay on Newtonian Matter-Theory and the Development of Chemistry*. Cambridge: Cambridge University Press, 1970.

Thorndike, Lynn. *A History of Magic and Experimental Science*. 8 vols. New York: Columbia University Press, 1923–58.

Trevisani, Francesco. "'Ratio' und 'Experimentum': Johannes Bohn (1640–1718) und die Italienische Experimental Physiologie." *Clio Medica* 17 (1983): 199–206.

Trevor-Roper, Hugh. "The Paracelsian Movement." In *Renaissance Essays*, 149–99. London: Secker and Warburg, 1985.

Turnbull, G. H. "Peter Stahl, the First Public Teacher of Chemistry at Oxford." *Annals of Science* 9 (1953): 265–70.

Ultee, Maarten. "The Politics of Professorial Appointment at Leiden, 1709." *History of Universities* 9 (1990): 167–94.

Underwood, E. Ashworth. *Boerhaave's Men at Leyden and After*. Edinburgh: Edinburgh University Press, 1977.

———. "Franciscus Sylvius and His Iatrochemical School." *Endeavour* 31 (1972): 73–76.

Van Berkel, Klass. "The Lagacy of Stevin: A Chronological Narrative." In *A History of Science in the Netherlands: Survey, Themes, and Reference*, edited by Klass van Berkel, Ablert van Helden, and Lodewijk Palm, 3–237. Leiden: Brill, 1999.

Van der Star, Pieter, ed. *Fahrenheit's Letters to Leibniz and Boerhaave*. Amsterdam: Rodopi, 1983.

Van der Wall, Ernestine. "The *Tractatus Theologico-Politicus* and Dutch Calvinism, 1670–1700." In *Spinoza's Philosophy of Religion*, edited by H. d. Dijn, F. Mignini, and R. v. Rooden, 2:201–66. Würtzburg: Königshausen and Neumann, 1996.

Van Helden, Anne C. "Theory and Practice in Air-Pump Construction: The Cooperation between Willem Jacob 's Gravesande and Jan van Musschenbroek." *Annals of Science* 51 (1994): 477–95.

Van Helmont, Joan Baptista. *Ortus Medicae*. Amsterdam: Ludovicum Elzevirium, 1648.

Van Leersum, E. C. "How Did Boerhaave Speak?" *Janus* 17 (1912): 145–50.

———. "Two of Boerhaave's Lecture Lists." *Janus* 24 (1919): 115–24.

Vanpaemel, G. "Rohault's *Traité de physique* and the Teaching of Cartesian Physics." *Janus* 71 (1984): 31–40.

Van Poelgeest, L. (1990) "The Stadholder-King William III and the University of Leiden." In *Fabrics and Fabrications: The Myth and Making of William and Mary*, edited by Paul Hoftijzer and C. C. Barfoot, 97–134. Amsterdam: Rodopi.

Van Slee, J. C. *De Rijnsburger Collegianten*. Utrecht, 1980.

Van Spronsen, J. W. "The Beginning of Chemistry." In *Leiden University in the Seventeenth Century*, edited by Lunsingh Scheurleer and Posthumus Meyjes, 329–43. Leiden: Universitaire Pers Leiden/Brill, 1975.

Van Swieten, Gerard. *Commentaria in Omnes Aphorismos Hermanni Boerhaave de Cognoscendis, et Curandis Morbis*. 6 vols. Venice, 1775.

Verbeek, Theo. *Descartes and the Dutch: Early Reactions to Cartesian Philosophy*. Carbondale: Southern Illinois University Press, 1992.

Vermij, Rienk. "Bijdrage tot de Bio-Bibliografie van Johannes Hudde." *Gewina* 18 (1995): 25–35.

———. *The Calvinist Copernicans: The Reception of the New Astronomy in the Dutch Republic, 1575–1750*. Amsterdam: Koninklijke Nederlandse Akademie van Wetenschappen, 2002.

———. "Mathematics at Leiden: Stevin, Snellius, Scaliger." In *Der "mathematicus": Zur Entwicklung und Bedeutung einer neuen Berufsgruppe in der Zeit Gerhard Mercators*, edited by Irmgard Hantsche, 75–92. Duisburger Mercator-Studien, vol. 4. Bochum: Brockmeyer, 1996.

Vesalius, Andreas. *Opera Omnia Anatomica et Chirurgica*. 2 vols. Edited by Herman Boerhaave and Bernard Siegfried Albinus. Leiden: Joannes du Vivie, and Joannes & Herman Verbeek, 1725.

Vigani, Johanne Francisco. *Medulla Chymiae*. Edited by David Stam. Leiden, 1693.

Von Haller, Albrecht. *Albrecht Hallers Tagebücher nach Deutchland, Holland, und England (1723–1727)*. Druck: F. Schwald/St. Gallen, 1948.

Wall, Faith. "Inventing Diagnosis: Theophilis' '*De Urino*' in the Classroom." *Dynamis* 20 (2000): 31–71.

Wallace, William A. *Causality and Scientific Explanation*. Vol. 1. Ann Arbor: University of Michigan Press, 1972.

Wear, Andrew. "Medical Practice in Late Seventeenth- and Early Eighteenth-Century England: Continuity and Union." In *The Medical Revolution of the Seventeenth Century*, edited by Roger French and Andrew Wear, 294–320. New York: Cambridge University Press, 1989.

———. "William Harvey and the 'Way of the Anatomists.'" *History of Science* 21 1983: 223–49.

Weber, Max. *The Protestant Ethic and the Spirit of Capitalism*. Translated by Talcott Parsons. London: Harper Collins, 1991.

Webster, Charles. *The Great Instauration: Science, Medicine, and Reform*. London: Duckworth, 1975.

Wiesenfeldt, Gerhard. *Leerer Raum in Minervas Haus: Experimentelle Naturlehere an der Universität Leiden, 1675–1715*. Amsterdam: Koninkijke Nederlandse Akademie van Wetenschappen, 2002.

Wightman, W. P. D. "*Quid sit Methodus?* 'Method' in Sixteenth Century Medical Teaching and 'Discovery.'" *Journal of the History of Medicine* 19 (1964): 350–76.

———. "William Cullen and the Teaching of Chemistry (I)." *Annals of Science* 11 (1955): 154–65.

———. "William Cullen and the Teaching of Chemistry (II)." *Annals of Science* 12 (1956): 192–205.

[Wolff, Christian.] "Relatio de Novo Barometrorum & Thermometrorum Concordium Genere." *Acta Eruditorum* (1714): 380–81.

Wolfganck, Isaacus. *Disputatio Chymico-Medica Inauguralis de Salibus*. Leiden: Abraham Elzevier, 1706.

Wolter, J. J. *De Leidse Universiteit in verleden en heden*. Leiden: Universitaire Pers Leiden, 1965.

———. "Introduction." In *Leiden University in the Seventeenth Century*, edited by Lunsingh Scheurleer and Posthumus Meyjes, 1–19. Leiden: Universitaire Pers Leiden/Brill, 1975.

Wright, John P. "Boerhaave on Minds, Human Beings, and Mental Diseases." *Studies in Eighteenth-Century Culture* 20 (1990): 289–302.

Yeo, Richard. *Encyclopaedic Visions: Scientific Dictionaries and Enlightenment Culture*. Cambridge: Cambridge University Press, 2001.

———. "Organizing the Sciences." In *The Cambridge History of Science*, vol. 4, edited by Roy Porter, 241–66. Cambridge: Cambridge University Press, 2003.

Zacher, O. "Die Bedeutung der Holländer in der ältesten Geschichte der Chemie." *Janus* 17 (1912): 535–56.

Zumthor, Paul. *Daily Life in Rembrandt's Holland*. Translated by Simon Watson Taylor. Stanford: Stanford University Press, 1994.

Index